"David Gessner is perennially p. [illegible] this fine volume, which asks us to actually think about and feel the world we are creating. It is an act of generational love and courage."

—BILL MCKIBBEN, author of *The End of Nature*

"Excellent environmental journalism."

—*KIRKUS REVIEWS*

"In A Traveler's Guide to the End of the World, David Gessner, a self-described 'polygamist of place,' bears powerful witness to the places he loves best, tracing the connections among their crises and finding possibility in their uncertain futures."

—MICHELLE NIJHUIS, author of *Beloved Beasts*

"With elements of dark humor that come flying like swallows going home, and with much beauty of detail, Gessner's journey becomes one of hoping to remember what's gone, writing a field guide to the life still present, as the reverent, remembering spirit of one human being."

—LINDA HOGAN, author of *A History of Kindness*

"This is a work of astonishing and visionary scope but also sharply intimate and grounded detail. David Gessner's kaleidoscopic journey sweeps in mammoth forces of nature, seemingly uncontrollable forces in society and economy, and an utterly refreshing, almost heartbreaking faith in language, communication and the potential of the human word to save the human world."

—CONGRESSMAN JAMIE RASKIN

"'Won't be water, but fire next time,' goes the old apocalyptic Black spiritual that my grandmother used to scare me witless with when I was a kid. David Gessner's 'Guide,' it turns out, brings her soothsayer singing close in, through the intimacy of friends and family. It is a different kind of reminder of our environmental predicament. A highly personalized accounting. I became a magpie peering over his literary shoulder, wondering if the usable air and water would run out before I could finish winding my way through the enraptured text. Though warmer and dirtier, it didn't and I'm happy for it, because his work is important reading. And he and my grandmother would disagree on one important point, it's gonna be water and fire, in our eventual end. The world will be alright, without us."

—J. DREW LANHAM, author of *The Home Place: Memoirs of a Colored Man's Love Affair with Nature*

"Gessner astutely tracks, guides, and plunges headfirst into the reality of climate change across our country in this visionary and crystalline portrait of how our world, our landscapes, and perhaps most importantly– our hearts, are forever altered."

—AIMEE NEZHUKUMATATHIL, author of *World of Wonders: In Praise of Fireflies, Whale Sharks, and Other Astonishments*

"Urgent but not panicked, this book is as personal and vulnerable as the world Gessner is describing. There are few better nature writers working these days, and even fewer who can convey both anger and possibility for the future in such a refined way."

—BRAD COSTA, Boulder Book Store

"David Gessner comes in hot with an indispensable contribution to our urgent conversation about how we should respond to the unprecedented environmental destruction our species has created. But this book is neither a bitter diatribe nor a dreary elegy. It is an engaging, entertaining, informative, unblinkingly honest look at the new world that climate change has wrought. Part poignant memoir, part adventurous travelogue, part accessible environmental science, this profound love letter to an uncertain future will forever change the way you imagine resilience, resistance, and transformation in the face of a rapidly changing planet."

—MICHAEL P. BRANCH, author of *On the Trail of the Jackalope* and *Raising Wild*

"The Clash was called 'the only band that matters.' When it comes to climate, Gessner is the only writer who matters. Others write book reports on global warming, spewing statistics. Gessner immerses himself, gets to know the people most affected while telling their stories, writing from inside the crisis."

—MARK SPITZER, Author of *Return of the Gar*

"In a sort of culmination of his writings to date, David Gessner invites us along on his journey to the end of the world as we know it. Visiting old friends and reacquainting himself with old (and very much changed) landmarks we see through his eyes, not just the changes wrought by current climate change but what happened in places like Chaco Canyon, Phoenix, and the Outer Banks hundreds of years ago. Yes, we've surely made a mess of things and yet, Gessner shows some possible ways forward and introduces us to people who are doing remarkable work. With his signature humor, Gessner manages to show us the worst while helping us hope for the best. Share this with your climate denial friends this year!"

—ANNE HOLMAN, The King's English Bookshop

PRAISE FOR DAVID GESSNER

"A master of the art of telling humorous and thought-provoking narratives about unusual people in out-of-the way-places."

—*The San Francisco Chronicle*

"A master essayist."

—*Booklist*

"For nature-writing enthusiasts, Gessner needs no introduction. His books and essays have in many ways redefined what it means to write about the natural world, coaxing the genre from a staid, sometimes wonky practice to one that is lively and often raucous."

—*Washington Post.*

"Contrary to the prevalent image of Thoreau the unsocialized, intolerant loner, the figure emerging from Mr. Gessner's book is, like Mr. Gessner himself, complexly alive, passionately in love with being on this planet."

—*The Wall Street Journal*

"I ended up spending more time in the company of Gessner's latest work than I have with any book I've ever read, save one: Walden. That is the highest compliment I could pay any book or its writer."

—*Washington Independent Review of Books*

A TRAVELER'S GUIDE TO THE END OF THE WORLD

For Orrin Pilkey, Fighter.

CONTENTS

PART IV.
A NEW WORLD

PART I.

WHERE THE ARROWS POINT

"A bird doesn't sing because it has an answer,
it sings because it has a song."

—Maya Angelou

YOUR TOUR GUIDE

➤ Let me take you on a tour of the end of the world.

No, no, you might say. *I don't want that: too depressing—I have my life to live, after all. My things-to-do.*

Well, consider taking a short break. Consider coming with me to places both beautiful and threatened. Yes, there will be more than a few dire facts about our warming world, and I will admit that during the year and a half of travel that constitutes this book, beginning with a visit to one abandoned capitol on the East Coast and culminating with a visit to another in the desert West, records were set for earliest storms, largest fires, latest first snowfall, hottest days, and an overall pace of catastrophe that makes your head spin, but I will also insist that, despite everything, our journey can still be kind of fun. Not *hopeful* mind you, that treacherous word I have come to regard warily, but *fun*.

How can that be, you may ask, *given the times?* I'm not sure. What can I say? It's a strange and sloppy world.

But facing that world, and not running from it, has this virtue: it is honest. And I, and maybe you, want to see and try to remember this still-beautiful planet. Its trees, its birds, its waters, its dirt and grass. This is a core difference between an activist's makeup and an artist's. Of course I hope we will stop devouring fossil fuel like drunken gluttons and of course I hope we avoid the worst consequences of climate change. But at heart I am not as interested in saving the world as I am in singing it.

↣ And yes I know that "end of the world" is overstatement. Life will go on in some form or another. Many will die but others will adapt. In the past I have resisted the apocalyptic. It seems grandiose. I once wrote a book that specifically objected to other books with titles that began with "The End of..." or "The Death of..." But, *what if*? What if we have really entered a new world? What if there is no hopeful plot twist at the end?

You will perhaps at least admit this: life everywhere is suddenly more primal. Right? It is taking more and more work to ignore the fact that we seem to be in the midst of an elemental comeuppance. During my travels I kept having the strange sense that I was living in the future, the same future that was predicted by scientists when I was younger but one that has arrived much faster than many of us expected. Time is strange; *then* becomes *now*. After years of debating climate change, we are inside it. Like many people of my generation, I first came to the idea of an altered future theoretically, through books like *The End of Nature* and films like *An Inconvenient Truth*. But while we may have already been at the end of nature, it always seemed to me that there was plenty of nature left, and back then it felt like they were talking about a time that was far away. It turned out we were wrong. It wasn't far away. We are in it now. For my daughter and for many of my students, there is nothing theoretical about facing a world where the elements—fire, water, wind—have turned against us.

↣ This book will concern itself mostly with the United States, where I happened to be born. It is also where I still live and where I have done most of my environmental reporting over the last thirty years. Though the 1.38 billion people in China consume more fossil fuels than any other country, the US still leads the league in consumption per person, at almost double China's rate. And while this book might be US-centric, it doesn't take much imagination to extrapolate. What I say about the sinking delta in Louisiana is true of the Bengal and Mekong Deltas on a vastly larger scale, and what is happening to the Colorado River, which quenches the thirst of almost forty million people, is happening with the Ganges and Indus and Mekong and the rest of the rivers that quench (or try to quench) the thirst of billions. So much depends on snowmelt. Just sub in the Himalayas for the Rockies. It might not seem like it at times, but in the end we are all in this together.

We have always had warnings of doom and always had prophets doing the warning. The Bible is chock-full of them. I am not a prophet, and as we travel together I will try to keep my proselytizing to a minimum. But I can't make any promises. Like the rest of my nature-writing ilk, I still like a good sermon now and then.

↣ During the year and a half of travel that makes up this book, I witnessed, without trying very hard, a flashflood ripping down a valley in Utah, the largest fire in the American West's worst fire season ever, the destructive aftermath of the fifth-strongest hurricane to ever make landfall in the United States, and a historic heat wave in the West that rivalled the Great Drought, which altered civilizations seven hundred years ago. The so-called disasters came so fast and furious that there was barely time to take a breath. During my journeys I experienced something that I have heard many people voice: Climate change has finally come home. It is not coming. It is here.

We have long written about global warming in one way: as a warning. But maybe we should now acknowledge that it is better thought of as a warning unheeded, like, say, Churchill's warnings about Germany before World War II. It is too late to warn about the coming war—the war has started, the bombs are dropping. We had better arm ourselves.

Speaking of Churchill, those who work in the climate-writing game might consider thinking about a sentence Edward R. Murrow wrote about him after his "We shall fight on the beaches" speech (one attributed at the end of the recent movie version of his life to Lord Halifax):

"He mobilized the English language and sent it into battle."

Maybe that is exactly what writers need to do when writing about climate. Try to use our words to wake people up. In this light I see my task ahead pretty clearly. Even if there is little chance of succeeding, I will attempt to tell the stories of global warming and to write vividly about what is happening to our world. Futile though it may be, I will try, as best I can, to mobilize nature writing and send it into battle.

↠ But even as I type those words they seem too gung ho, too military. Lately there's been a lot of talk about *weaponizing* this or that. But climate change is not war, even if the results will eventually be much worse than any war. More modestly I will say this: we need a new way to tell this story.

What to *make* of the end of the world? That is one of my guiding questions. It is a writer's question. Another question that I find pressing upon me is this: how will people look back on the literature of our time if it does not address our major existential issue?

By literature I don't mean propaganda, nor do I mean fact-spewing book reports that read like television punditry frozen into print. In most of the writing about climate the sentences are not sloppy enough, too uncomplicated. And they are not BIG enough. The language does not rise to the challenge. Literature is not policy.

I know what doesn't work. A list of horrors and statistics. Doom alone doesn't inspire. Perhaps the modest and long-scorned genre of nature writing can help. I am not naïve and, given the scope of the crisis, I don't see a gang of nature writers riding to the rescue. But on the other hand, it is a genre well-suited for contemplating the world beyond the human and how we interact with that world. By having first-person encounters, not alone but with other people and animals and habitats in the climate-plagued places that increasingly make up our world, we can help make the larger story personal.

➤ "Each disaster is that person's experience," a victim of the Paradise fire said to me.

Exactly.

FUTURE AIR

→ My daughter, Hadley, is nineteen years old. I am sixty-one. (Yikes.)

→ Not long ago I read a description of the future of air in a book called *The Future We Choose*, written by two of the architects of the Paris Climate Agreement. The authors, Christiana Figueres and Tom Rivett-Carnac, are not doomsayers but they have written a vivid description of what the world will be like in 2050 (when Hadley will be in her late forties) *if* we don't take serious and bold action:

> The first thing that hits you is the air.
>
> In many places around the world, the air is hot, heavy, and depending on the day, clogged with particulate pollution. Your eyes often water. Your cough never seems to disappear...You can no longer simply walk out your front door and breathe fresh air: there might not be any.

According to the authors, one thing that won't be going away in the future will be face masks. We'll just have to repurpose them.

→ There was a question looming over my climate explorations, a question for all of us, one that adds some suspense to the proceedings:

Can we—inept, contradictory, self-interested creatures led by compromised, sluggish, corrupt governments—change?

Will we change?

Stay tuned.

Sometime during my year and a half of travel I whittled that larger question down into something much more manageable and specific, something that I decided to ask scientists, environmental thinkers, and pretty much everyone else I ran into. The question was this:

What will the world be like in forty-two years?

The question is a father's question.

I have promised myself not to proselytize, to try to describe the world "as it is." But of course "as it is" is changing. No one imagines this is the end of nature's fury. I have also used the phrase "living in the future" to refer to what life feels like now. But what about the actual future? The future of weather? The future of heat? The future of storms? The future of fire? The future of human beings trying to adapt? The future of community and commitment to place? The future of, god help us, government? And what will climate adaptation look like? Will it be worse or better than we imagine?

These are the questions I began to pose to those who know more than me, and I wanted to be specific. Why forty-two years? Well, the idea for this book grew partly from my daughter's anxiety about the future, so I naturally began to wonder what the world would be like when she was my age. Sure, I could round it up and ask what the world would be like in fifty years. But that's the normal question, the boring question. I decided to get more exact. I was forty-two years old when Hadley was born, so I asked scientists and thinkers to paint a picture for me of what the world will be like in 2063, when she will be sixty, the age I was when I began to travel again in 2021. What exactly will she be facing?

The year 2063 is not that far away. Think of how time moves. I remember being eighteen and graduating from high school like it was a blink ago—1979. Now flip that into the future. What will it be like? *Really* like, not apocalyptic-book-like or Fox-News-like or techno-fix-Jetsons-like? Will we be able to step outside in the summers? Will we be able to see the stars at night? Will the fires have burned the West black and will the lifeless seas have drowned the cities along the Atlantic coast? Could that *really* happen? I wanted to hear honest, put-your-money-down predictions. I was not looking for worst- or best-case scenarios or being saved by technology or massive carbon reductions but a cold-eyed assessment of where we as a species, and where Hadley as an individual, will really be.

It is a time when I will not be alive.

So what will that world be like? And how will my daughter fare in it?

➤ I'll admit it was kind of fun to give prominent scientists a creative writing assignment. A lot of them chickened out. They are cautious folk, after all. One who didn't was Caltech's Paul Wennberg, the R. Stanton Avery Professor of Atmospheric Chemistry and Environmental Science and Engineering, who studies the influence of human activity on the global atmosphere.

He wrote: "For those old enough to remember, the sunsets in the early 2060s are reminiscent of the year after Mt. Pinatubo erupted in 1991—the deep purples and reds and a sense of seeing the sky long after the sun has gone down."

That's a pretty good start, right? Not bad for a scientist. I'd give him an A.

➤ "Doom is normal," Hadley said to me the other day.

It was June of 2022 and we were renting a cabin in Boulder, Colorado, and sitting at the picnic table in the backyard in the shade of two apple trees. Magpies yammered down at us. I had just been freed, the five allotted days passed, but to be safe we sat at opposite ends of the table and I wore a mask.

The magpies never shut up. I don't blame them, of course, it's their nature, and that's something we share. Though I spend many hours each day alone making sentences, I am also relentlessly social, which means that during my first three days back here in Colorado, before I realized I was sick, I hugged at least a dozen old friends hello, friends who I later had to write to tell I tested positive.

We are one of the lucky families who avoided COVID-19 for the first two years of the pandemic. Until this very week in Boulder, when I finally succumbed. So I spent the better part of the first five days of our vacation in solitary confinement in the cabin bedroom, speaking to Hadley, my teenage daughter, through my bedroom window while she stood outside, keeping back ten feet. I had no television while in lockup, but the television-sized square of the window next to my bed became my chief entertainment. (To clarify I mean television-sized circa 1990.) The show going on outside was nature on a small scale and the main characters were the magpies. Big birds with flashing white patches and resplendent blue wings and black backs. I watched as they fished for slugs in the cranial dip where the rainwater collected in the roseate-colored rock by my window, or as they flew from the apple tree to the banana yucca, or hunted for worms and fed their young.

My wife and I met and married in Boulder over twenty-five years ago, and we still have a great group of friends here. With COVID I missed the usual frenetic pace and raucous feel of our trips back, but there were benefits to my isolation. I spent the quiet days mulling, and trying to make sense of, my travels around the country, travels that began with a trip to Washington, DC, in March of 2021, about a year after the pandemic began.

➤ Some people still insist that those who worry about climate are overreacting. That things have not changed much. The earth is still the earth. Life is still just life.

Maybe they are right. Maybe we are not really in an age of crisis. Maybe the environmental scaredy-cats are exaggerating. Maybe it will just blow over.

Maybe. But if I were forced to debate this notion, I would present Hadley's high school years as Exhibit A.

Consider:

During the fall of her freshman year our family evacuated due to Hurricane Florence and her high school was converted into a shelter for the storm's victims. There was no school for close to a month. The next fall Hurricane Dorian hit, closing school again, and the spring of her sophomore year the pandemic struck. All of this punctuated by the now *de rigueur* bomb scares and shooter warnings. Spring freshman year was her single disaster-free term. On the bright side, she had to greet only one hurricane while wearing a mask.

➤ Dr. Wennberg, my first scientist respondent, continued his foray into climate fiction:

> By 2050, even though CO_2 concentrations were now close to stabilizing at 500 ppm, methane concentrations continued to rise and the Earth was simply too hot. Summers in much of the subtropics were literally unbearable; droughts and fires had spread out far from the usual places (US west, Portugal, Australia) and were causing huge property losses across the world. Repeated crop failures were causing famine.
>
> Through the United Nations Environment Programme (UNEP), the vast majority of nations had a decade earlier approved a scheme to inject sulfur into the stratosphere to reduce global temperatures back to those of 2030. Specially designed planes flew daily into the lower stratosphere delivering H_2S together with nucleating particles designed to produce a nearly uni-

form stratospheric haze capable of reflecting 1% of the incoming sunlight back to space and knocking almost one degree off global mean temperatures. In approving the scheme, the UNEP received commitments from all the signatories to pay for atmospheric CO_2 removal. Most nations had chosen to use enhanced ocean alkalinity efforts and were mining limestone, milling it to small particles, and dumping it across the world's oceans. It appears to be working and scientists now predict that CO_2 levels will decline back to 420 ppm in the next 30 years. If successful, by 2100 the sulfur Band-Aid will be able to be pulled off completely.

Dr. Wennberg's description of how we will respond to catastrophe seems about right to me. Even if parts of the world can break their fossil fuel addiction, other parts will understandably cling to it, wanting what we have, or, by then, *had*. And while we are currently debating atmosphere-altering technologies, is there really any doubt, given our human tendency to meddle and fix, that we will embrace them if we find ourselves in a burning world?

➣ A recent Gallup poll says that only 3 percent of Americans believe that climate change is the most important problem the country faces.

➣ The magpies have been the presiding presence around the cabin we have rented at the uppermost point of Chautauqua Park, in the shadow of the Flatirons in Boulder. The cabin is less than fifty yards from the one I lived in for two years in grad school thirty years ago, so I can look out my little window and see my past. Hummingbirds, western bluebirds, and towhees populate the nearby trail where I took my coffee this morning, but the gangs of magpies rule the neighborhood. One of the few species that give humans a run for their money in self-awareness and sheer verbosity.

I thought it would be apt to conduct the interview with my daughter outside, at this picnic table not far from the burn scar of the recent fire in the Boulder foothills, but the magpies have other ideas. They will not be ignored. So rather than ask Hadley about climate change, my first question is what she thinks of the birds.

"They're funny, they're loud, they're obnoxious. But they're pretty."

That's about right. Like other corvids, including crows and ravens and nutcrackers, they are relentlessly curious, smart, and vocal. And if you can manage to ignore their pushy personalities you can lose yourself in the oil spill blues, greens, and purples of their backs and tails.

➤ To imagine the future, consider the present. Currently seven million people a year die from air pollution.

➤ Before Hadley came back from a hike to join me at the picnic table, I had been sitting here wondering how I would tell her that a rabbit had died and was lying below the tree of heaven that grew beneath her bedroom window. Then, just before she returned, the rabbit popped its head up, shook itself off, and hopped away. Another example of my questionable skills as a naturalist. The bunny was just napping.

"I used to talk about it a lot more," my daughter now says as we sit at the picnic table.

The "it" is climate change.

Hadley seems to have been born with an activist gene that I lack. And since I have been running around the country interviewing everybody else about global warming, it is high time to talk to my teenage daughter.

It takes a while for the magpies to quiet down enough for us to continue. When they do I ask about her origins as a climate activist.

"I became climate conscious in eighth grade when I began connecting the dots and realizing that this was not good for my future. I felt a sense of obligation because almost no one I knew at middle school really seemed to care about climate because their futures were all planned out and paid for, but I was worried. Then in 2019, my sophomore year of high school, I really started to get involved. That was the year I became a vegan, spoke at a climate rally at city hall downtown, and started a Sunrise group that met every month."

Sunrise is the youth wing of 350.org, the organization that climate activist Bill McKibben founded.

"Spring of 2019 was my last normal year of high school. Then the hurricane and COVID hit and put a wrench in my whole activist plan. My friends and I had a feeling of like 'Oh shit, this is *it*, it's happening *now*, there's nothing we can do.' It was too big for us. I couldn't speak up any longer except through the internet. We kept the Sunrise meetings going online for a while but then quit. It was really exhausting because of the state of the world. It felt like my activism wasn't doing anything and I wasn't capable of really changing anything."

I considered this: by seventeen my daughter was a disillusioned activist.

"It just got so stressful and exhausting to put the idea of our futures in the forefront. Just thinking about it and my own potential future being not so great because of climate change was all consuming. Which was not so fun. And while I still wanted to participate in activism, I also wanted to have a decent high school experience, and to sometimes set it aside and focus on life, existing like a normal person."

⇥ If you really want to imagine the air your children will be breathing you could do worse than reading *The Ministry for the*

Future. In this work of fiction, written by Kim Stanley Robinson in the style of the sci-fi novels that made him famous, Robinson describes the unending drought in India that takes place not in 2063 but in 2025:

> It is too hot to cough; sucking back in air was like breathing in a furnace, so that one coughed again. Between the intake of steamy air and the effort of coughing, one ended up hotter than ever.

➜ Last year *The Lancet*, one of the oldest and most-respected general medical journals, published a study of ten thousand young people, ages sixteen to twenty-five in ten countries, which revealed that the majority of the respondents experienced climate anxiety as a regular part of their lives. The study concluded: "Distress about climate change is associated with young people perceiving that they have no future, that humanity is doomed, and that governments are failing to respond adequately, and with feelings of betrayal and abandonment by governments and adults. Climate change and government inaction are chronic stressors that could have considerable, long-lasting, and incremental negative implications for the mental health of children and young people."

➜ When I asked Hadley if thinking about the climate crisis left her feeling sad she said no.

"The sadness doesn't come through as much anymore as the anger does. I can't mope. Or I mean I try not to mope. The main emotion I feel is anger at the people who did this. There are people who could fix this, people with money and power, people who could start to solve this and they're not. And that is what makes me mad."

I have been on earth for six decades now, which means I have gotten pretty good at repressing, at pushing the bad stuff down.

If my daughter's brain is still developing, mine is going the other way. I can't really feel what she feels. But listening to her I am angry too.

We have failed our children. That seems obvious enough. Given all the evidence, we have failed to imagine the future and act on what we have imagined. It is, among other things, a massive failure of empathy.

I wonder: Are we really so empathetically challenged that we can't see the mess we are leaving behind? Are we a bunch of drunken frat boys who have decided, what the hell, we might as well trash the place? I know I am culpable; I am one of the mess-makers.

To imagine the lives of those who will come after you. It is one of the essential imaginative acts. Picturing the lives of our children's children's children. But since we are too imaginatively stunted, let's not even go that far. Let's stick with one generation. Can we at least do that? Can we imagine the lives of our children?

➤ And yet it is complicated, right?

"Climate change is not at the forefront of my brain," my daughter says. "I am still a nineteen-year-old girl."

Maybe she, like 97 percent of the population, like you perhaps, is sick of hearing about the climate crisis. Maybe she is ready for her father to start working on a different book on a different, happier subject.

I am not a nineteen-year-old girl but, like my daughter, climate change isn't always in the forefront of my brain. We all have multiple lives and slide between them, sometimes incongruously and awkwardly, sometimes easily. It is true that the seas may rise and drown the East Coast and fires burn the West. But how can that compare with the promotion we might get at work or the fight we had with our spouse or the moods of our teenage daughters?

And yet.

"When I'm alone and really think about it, it freaks me out," Hadley admits.

➤ Let's end where we started. Not with the usual images of rising seas and burning forests but with something simpler. With *air*. Try for a minute to imagine it, to really imagine it. You stay inside the house because to step outside means to breathe in the acidic taste, and you never feel like your eyes will stop burning or your throat stop scratching. Not long ago I went walking through a burnt forest after a prescribed burn. A black landscape of ash with logs still smoking, burning from the inside. I found myself coughing and choking, not quite able to take a full breath.

That is the future of air.

NATURE WRITING BY THE NUMBERS

GESSNER '09

→ My faith in number six is wavering.

FIRE AND WATER

➤ This morning while I write outside at the picnic table a magpie is yakking down at me from the buckthorn tree. I retreat into the house for a bit and when I come back another magpie is poking its head into my water glass on the picnic table. Less annoyingly, I later watch as a third bird, sitting up in the apple tree, feeds a worm to one of her brood. The youngster is already the size of a crow.

Now that I have been officially cleared to re-enter the world, my wife, Nina, and I decide to hike up above Boulder to see the burn scar. We follow the Bear Canyon trail as it turns into a dirt road and heads into the mountains. The NCAR Fire, which started on March 26, burned just shy of two hundred acres and forced thousands of evacuations. This year saw three significant fires in Boulder County.

We have been living in North Carolina for nineteen years and the whole time we have fantasized about moving back to Boulder, despite the prohibitive cost. But when we talked about it again this spring, I was surprised that Nina, who had never wanted to leave Colorado in the first place, said she didn't like the idea of going back. Why? It was too dangerous.

➤ A close friend in Boulder sends out one of those holiday letters each year catching us up on his and his family's lives, and the letters are, as the genre requires, mainly hopeful. Sure enough, this year's letter was sprinkled with reasons to be thankful. But as he said to me on the phone: "It was not a good year for Boulder County." He was not talking about COVID.

Here is the opening paragraph of his holiday wrap-up:

> Last week, for the second time in 2021, we received texts from friends and family around the country asking if we were safe. A wildfire (*in late December*), sparked by a downed electric wire and fueled by 100 mph winds and historically dry conditions, burnt down close to 1000 houses in our country...We know of at least four families that lost their homes and several others whose houses barely escaped being destroyed. The first tragedy was in March, when a man armed with a semiautomatic rifle randomly killed 10 people in our neighborhood grocery store.

He added: "We don't feel as safe as we used to."

➤ In this new age there are fewer and fewer places we can think of as safe. The anxiety that Hadley describes is not confined to teenagers. It is not just a western feeling either. Living as I do on the Carolina coast, and traveling the West during the summers, I have long been struck by the way that eastern hurricanes and western wildfires mimic each other. I have found that people use the same apprehensive language as the fire or hurricane seasons (coming ever earlier) approach. There is a lurking not-always-conscious sense of dread. As we hike up to the ridge my wife and I are well aware that when we head home hurricane season will be underway, and that the old rules for fire season are off.

➤ The second reply to my query about 2063 hit even closer to home. Dr. Wennberg's answer described the world burning up. Orrin Pilkey's described a flooded world. Orrin is a coastal geologist and emeritus professor from Duke, a controversial figure in coastal studies, who over the years has become my good friend. The flooded world he described was my world.

When people ask me how long I have lived in coastal North Carolina it isn't hard to calculate the answer. Hadley's age serves as a mnemonic device since we moved from Boston to Wilmington, on the state's southeast coast, when she was three months old. It didn't take long to understand that wind and water behaved differently in this new place where we found ourselves. We quickly learned that the rising sea was no abstraction in our new home.

Orrin Pilkey and I first met after Hurricane Isabel, which hit soon after Nina, Hadley, and I moved to North Carolina in 2003. Isabel passed, but the idea of it lingered. While the storm did no serious damage to our new apartment, for me it served as a kind of primal wake-up call. A fact that I had been numb to came alive: more and more scientists believed that the actions of human beings, while not actually creating storms, were altering them, making them larger, longer, wetter, more intense. We all know this now, and many knew it then. But living on a barrier island it took on a new and pointed relevance. I began to read up on coastal geology and the history of storms.

The next year we evacuated again. The year after that, our third in the South, we would have to evacuate our home twice and it would prove to be the busiest hurricane season in memory. Before it was over three of the six strongest hurricanes ever recorded (at the time) would make landfall, and one of those storms was named Katrina.

In the course of my hurricane self-education I kept coming across one name again and again. Pilkey. Orrin Pilkey. He had long been a lightning rod in the coastal battles of North Carolina and beyond. Some people spoke of him fondly, others not so fondly, but everyone spoke of him. "An idiot with a beard," one local town planner called him. But from what I read it seemed clear that others saw him as a kind of prophet, fighting against over-development. In that role his main message, the one he had brought back from the shore as if carved on sand tablets, was a simple one: *retreat*. By that he meant we should retreat from the beaches, and that, rather than rebuild after storms, we should let the buildings fall into the sea. As you can imagine, this was not popular with homeowners, realtors, or the boosters at the chamber of commerce.

"The way to go, I think, is to relocate and get out of the way or stay and do nothing," he said to the homeowners. "If the buildings fall into the sea, they fall into the sea."

Orrin's answer to my question about the future focused on the Outer Banks of North Carolina. There are more barrier islands along the Atlantic and Gulf Coasts than anywhere else in the world, and the string that forms the Outer Banks of North Carolina is perhaps the most beautiful, fragile, and threatened of those islands. The islands of the Outer Banks sit off the state's coast like a fragile shield, and, when it comes to sea level rise, are the closest thing the East Coast has to Guam.

Orrin's sea level predictions are higher than most, seven feet by the end of the century, half that by the time Hadley is sixty, and if they come true the Outer Banks will be under water by 2063. But Orrin believes that barrier islands will not be developable with as little as a one-foot sea level rise. Well before they are inundated they will be sliced to pieces by future storms. According to Orrin's predictions, there will not be anyone living there when my daughter is sixty.

When he answered my question Orrin also pointed out a nice little irony: barrier islands were formed by sea level rise in earlier epochs. What people see as shoreline erosion is just the island attempting to migrate.

According to Orrin, several factors make the Outer Banks the most dangerous barrier island chain on the East Coast. They are narrow and low with a steep shoreface and are battered by the high wave energy of the Atlantic. They also have large bodies of water behind them to form inlets, and escape is either by ferry or on narrow two-lane roads with sandy shoulders. Given all this, Orrin says, "Modern dense development of barrier islands is madness."

And that, he concluded, is just the Outer Banks.

The coast behind them will also be submerged at least two miles inland.

Which means that it is unlikely that middle-aged Hadley will be coming home for any high school reunions.

➤ Nina and I reach the rock ridge. The ridge is a dividing line between two worlds. On one side a green forest, on the other black skeletal trees, some bare but some still holding onto their dead dull orange needles. Almost within reach is a fully blackened and burnt juniper, its roasted berries still visible. I have eaten these berries before but never cooked and I don't recommend them served either way. We stand atop the ridge on rocks stained nearly malachite green with lichen. A golden eagle soars above. I have seen video of the fire. Moving at the speed of a sprinting man, pushed by the wind. It was controlled just a couple hundred yards short of a line of houses. A small fire really, compared to the big one in late December.

Down below us, like a magnificent sandstone castle, sits the building that gave this fire its name. This is the National Center for Atmospheric Research (NCAR), where some of the world's top climate scientists had a front row seat at the exact sort of disaster they study.

One of those scientists, Ronnie Abolafia-Rosenzweig, is a postdoc who recently copublished a paper on creating a method for predicting seasonal wildfires based on analyzing precipitation, temperatures, drought, and other climate conditions in the winter and spring. Dr. Abolafia-Rosenzweig was also one of the respondents to my question of how Hadley would fare in the future. Not surprisingly, he confirmed what most people in the region know: to live in the West in 2063 will be to live in a land of fire.

He began with a short history, including a fact that more and more people in the West have come to understand over the last decade of conflagrations: much of the kindling stoking today's megafires accumulated during a century of "successful" fire suppression. That kindling consisted of all the unburnt material in the forests themselves but also the houses that were built in places that had historically been in the fire zone. Ironically, those houses, which would also serve as kindling for the megafires, were only built because fire had been "eliminated" in those places for so long. Fires, set historically by Native people or ignited by nature in the form of lightning strikes, had been intertwined with the evolution of the western landscape, but these new fires, fed by increased temperatures, a drought that has no end, and a surprisingly fragile power grid, are different. It is not that there weren't big, even bigger, fires in pre-suppression days, but they did not come one after another during a single season and they have never burned so hot.

"Western US fires are burning at higher rates and intensities than previously recorded in recent data records from satellites as well as millennia-long records from tree-ring data," Dr. Abolafia-Rosenzweig wrote me. "Some forests in the western US are at their highest density relative to multi-millennia-long paleo-records due to the legacy of fire suppression following Euro-merican colonization and Indigenous depopulation. Essentially, we have been stockpiling fuels while changing the atmospheric composition to favor hotter and drier (i.e., more flammable) conditions. Now we are experiencing the most active fire seasons in historical records."

And what about forty-two years from now?

I expect that over the next forty-two years, fire

seasons will start earlier and earlier and end later and later resulting in longer and drier seasons that become increasingly favorable for fire. These longer, hotter, and drier fire seasons will likely enable fire activity of never-seen-before severity.

Western US climate is projected to become increasingly conducive for fire. For instance, a recent study in Nature's *Communications Earth & Environment* found that the climate from 2021-2050 will be twice as conducive for forest fires relative to the prior thirty-year period (1991-2020). Climate and fire are very tightly coupled in the western US: with climate variability and trends explaining at least 70 percent of the year-to-year variability and the increasing trend in fire activity over the recent four decades. Thus, it is extremely likely to experience increasingly severe fire seasons over the next decades as warming increases. There is a massive amount of forested land to burn before fires become "fuel-limited" (i.e., there is no longer an abundance of fuel to burn), and thus it is expected that year-to-year variability in fire season severity will likely be largely dictated by climate over the next forty-two years.

➤ "There are lots of angry vegans on TikTok and Instagram," Hadley explains. "Early on I was part of an online community of vegans."

She remained a vegan, but has long since tired of the online groups.

"I didn't last long because so many of the people on it were so angry at everybody else. They didn't think to educate people instead of yelling at them.

"I don't want to yell at people. It's all tangled up. I don't think there is anything wrong with eating chicken nuggets and staying on your phone all day and not lifting a finger, though the second you learn these behaviors are bad—or not even bad but could be negatively affecting the world—maybe you could change. It's exhausting when people say they want to change but then get angry and say, 'Hey, don't push this on me.' But I don't think you can shame people into changing."

I tell her that lately it feels like everything that I once read and was predicted by scientists like James Hansen of NASA and by writers like Bill McKibben, scenarios that so-called skeptics mocked, are now coming to pass.

"Does it feel like things are closing in?" I ask my daughter.

"Things *have* closed in. Ever since I can remember, the hurricanes and fires. I have only existed when things have been bad. COVID was like a nail in the coffin. I can't remember a time when things were peachy keen. Yes it feels like impending doom, but the doom has always been impending. Like I told you, doom is normal."

Of course hers is not the first generation to feel this way, as anyone who had to hide under a desk during air raid drills or suffered from nuclear nightmares can attest. And despite lamenting that she was "never going to have a normal year of high school," I was a little envious of some aspects of my daughter's high school experience. I had been more of a loner during my high school years, while Hadley was outgoing and sunny despite the doom. Throughout those four years she stayed extremely close with a group of four other girls, a pod that, despite the pandemic, enjoyed sleepovers, parties, beach days, camping trips, and cross country meets. And yet it is true that she took most of her classes not at school but from her bedroom.

"I just think I come from an exhausted and depressed generation," she continues. "Almost every person I talk to these days has dealt with either exhaustion or depression or both. I don't sleep. I'm anxious and exhausted most of the time. Kids my age can either ignore the world or do something about it. Both options are exhaustion- and depression-inducing. You can't just lock yourself in your room all day if you want to be a happy person. I get it and I lose myself in my phone too, but distracting yourself with media is not going to be enough. Because eventually the end of the world is going to creep in."

➤ This may sound, in part, like the lament of teenagers from time immemorial. Except that these thoughts, which, like a teenager's brain, may not yet be fully formed, are born of experience. And while Hadley may have witnessed her high school transformed into an emergency shelter during Hurricane Florence, she has nothing on her best friend in Boulder. Hadley and Lucy met at art camp during one of our summer visits to Boulder when they were twelve and eleven respectively, and they have spent the last week together. Lucy actually lives outside of Boulder, in Louisville. Back at the end of December Lucy came to visit us for New Year's. Lucy had interacted with Hadley's high school friends from North Carolina so much online that they practically felt like her own classmates, and she was excited to meet them in person.

On December 31, I drove Hadley to the airport in Durham to pick up Lucy. Lucy had flown out that morning, and shortly after that her parents had flown to California on what would turn out to be one of the last planes out of Denver. Hundred-mile-an-hour winds were blowing over the mountains from the west. Those winds downed power lines that were originally blamed for sparking the fire that started just south of Boulder in the town of Marshall. Whatever started it, the flames, fed by drought-cured grasses, sped east toward Louisville and Lucy's house. Over the previous six months, Denver and the front range had gotten just one inch of precipitation—a record low for the second half of the year—and the area had seen the latest first snowfall in recorded history. With no snow at all and wild winds, southern Boulder County burst into flames.

While Hadley and I were picking Lucy up at the airport,

the fire was closing in on her home. Five hundred houses had burned by then.

Lucy knew about the fire, but whether she knew how close it had gotten to her house we were not sure, nor were we sure how much we should tell her. Lucy and Hadley went about the business of normal teenage life, attending a New Year's Eve party where Lucy got to make her virtual Carolina friends real. Meanwhile Nina and I followed on the computer and through texts with Boulder friends as the fire made its march toward Lucy's neighborhood.

"I don't know if we are equipped for this," Nina said when considering the prospect of telling our daughter's friend that her house had burned down.

It seemed to me that *I don't know if we are equipped for this* is a pretty good mantra for our times.

↠ Apologies.

I am not a very good Jeremiah. I do not have a fiery vision of the future to proclaim for you. I cannot scold you for the way you have squandered our bounty. I have squandered it too. I am very much a part of what a true Jeremiah would scorn. The oldest son of a big family of eaters, drinkers, consumers. A man who uses enough water and gasoline each day to be regarded a criminal in the future.

When the vision does start to come, cobbled together from the replies of scientists, my imagination quails at the picture they paint. How can I really imagine what Hadley will see? They say the air will not just be too hot but thick with particulate poisons. Air that will make the well sick, and cause the sick to die. The heat will be unbearable in cities and inland, where 100 degrees will be long since left behind, then 120, 130, the numbers creeping higher and higher. Birds, my joy and respite, will fly in low numbers so that our time now will be remembered as we currently remember the great migratory flocks of passenger pigeons. The ocean, devoid of life, will no longer be able to absorb the carbon dioxide we have poured into it, and the fires in the West, fed by the drought and lack of snow, will burn year-round. If the glaciers go then all bets are off: the cautious incremental predictions of scientists past will be laughed at. Orrin will seem cautious. Even if they don't melt, coastal cities and towns will flood, their infrastructure rendered useless. Salt water will be everywhere but fresh water will no longer be a readily available resource, will no longer flow from a tap, will be something that is paid for by the gallon or quart. In that it will be like all resources: available to the wealthy as the gap between rich and poor widens. Those who have will hoard and use. If this proves true can violence over water, over all the scarce resources, be avoided?

I know all this, you see, but I can't really picture it, can't really *feel* it, can't really imagine Hadley walking through it when she is my age. But the fact I can't imagine it doesn't mean it isn't going to happen.

I am better at describing the world as it was and is than as it will be. Maybe the life I have lived will be looked back on as if it were a dream. It is a lucky life where I have walked into national parks and forests and believed I have *gotten away*—what a concept—before Mr. McKibben declared the end of nature, back when I not only splashed creek water on my face but sometimes *drank* it, and when I did I felt exhilaration and I felt freedom and I felt joy. I remember seeing flocks of white pelicans flying overhead and seeing humpbacks breach from shore and seeing a grizzly try to rein in its frolicking young and seeing thousands of swallows stage in a great cyclone of birds and paddling alongside a dozen dolphins and swimming, always swimming, in cold, clean water. I remember doing this, in the East and West, in mountain streams and in the oceans, oceans that we are told will soon be hot pools of infection but were once our great respite, our great pleasure. I have been lucky to have these days upon days in nature without worry that the forests will be gone, burned, beetle-ridden.

And so even as I type these words describing what will become of the green world, I still can't quite completely believe them.

Could this really happen? Could this really be?

Maybe we all remain climate skeptics of a sort. If we can't imagine this will happen, if we can't picture it, maybe we should all be lumped in that category.

If we weren't, if we really *did* believe this future could happen, we would act in a way that made one thing clear: this is humanity's greatest priority. This is the most vital thing, the looming thing. The thing we are choosing not to face.

The challenge is to get beyond words. The challenge is to *feel* the future.

LOSING EVERYTHING

→ It is a writer's worst nightmare.

True, it is a human nightmare as well, but it resonates especially with those scribblers of words, keepers of journals, and hoarders of paper who call themselves writers.

Technically, Ken Sleight, ninety-one years old with a birthday on the way, has always been more of a character than a writer. Sleight was immortalized by Edward Abbey in his classic 1975 novel about eco-sabotage, *The Monkey Wrench Gang*, as Seldom Seen Smith, the "Jack Mormon" who broke away from his childhood religion and became a famed river runner known for fighting back against overdevelopment in the American West. Sleight, like his fictional counterpoint, had grown up conservative—he was a member of the John Birch Society as a young adult—but later began running rapids and blazing trails that others would follow. By the time he confronted developers on the Colorado Plateau, and rode his horse into a standoff with approaching bulldozers, he had evolved into a full-blown eco-hero.

For thirty-five years Ken and his wife, Jane, have lived and raised horses and goats up at Pack Creek, a paradise of green at the foot of the La Sal Mountains that looks down upon the orange redrock rim above Moab, Utah. The place is ideally perched: look one way and you see the cool green of the mountains, the other the dry, shimmering heat of the desert. During those years, Sleight's daily commute has been the quarter-mile walk from his home around the field that holds the goats, horses, and chickens up to what he called his shop, a large, two-story Quonset hut. Quonset huts, first developed by the US military during World War II, are lightweight, prefabricated structures of corrugated galvanized steel with a semicylindrical cross-sectional style. Ken's looked like half of a giant aluminum beer can jammed horizontally into the ground and made generic by the weather. Three times over the course of the last decade and a half I have visited Ken and every time I found him there, in his shop. Each visit he was basically in the same spot where I'd left him, as if he hadn't moved since I was last there. He would be sipping whiskey out of a coffee cup or nursing a beer, sitting behind a desk that was cluttered with papers, old computer discs, and two computers, one defunct or at least unplugged. Despite the heat, he always wore a flannel shirt with a blue dungaree shirt over it, and he always smiled widely upon my arrival. True, he looked a little older each time, but none the worse for the wear: shaggy white eyebrows, hunched shoulders, ears that stuck out like jug handles, a big thatch of white hair.

Boxes and files were everywhere in the shop. Records of all his early river trips, photos of the same, transcripts of his many environmental fights, personal memorabilia, minutes from the meetings of the San Juan County Democrats, notes for the book he hoped to write about his friendship with Edward Abbey. In short, all the paper that made up his life. It seemed like everyone else who had known Abbey had written about their experience, but Ken's book would be different: he told everyone that one day he would spin his notes into gold and tell the story of his early days as a river runner and a friend of Ed.

Whether he would have or not is beside the point. The point is that the project, never finished, always looming, gave his life a secret purpose. A sense of unfinished mission. If you have purpose, I've come to think, you can endure pain and loss. What was really stored in the Quonset hut, along with farm equipment, a woodstove, a never-used exercise bike, filing cabinets, and dozens of empty beer cans, was his past. Hundreds and hundreds of boxes of the words and pictures that constituted his memories.

The fire swept down the mountain first. The investigators later concluded that it had been the result of an "untended campfire" in the public picnic area above Pack Creek on June 9, 2021. This would result in much understandable outrage in the small community, but whatever its cause, the fire had its own agenda. Ken, Jane, and the rest of the members of the neighborhood were evacuated as the flames made their way down the north side of the creek. Ken, worried about his goats, unlatched the pasture door before being led away, and for the next two days he had no idea whether his home, or those of their neighbors, had survived the blaze. The early reports, which he heard while waiting at a hotel down in Moab, were not good: they could see orange flame lighting up the night sky and it was said as many as thirty houses were lost. Luckily those reports were exaggerated. The local firemen, fighting heroically door to door, saved every house but one on the north side of the creek. By the next morning the fire seemed to have stopped at the bridge over Pack Creek below the house. The next day federal authorities claimed that it was under control.

They were wrong. Most of us live with some illusion of control, but one thing we know we can't control is the wind. The wind had pushed the fire down the creek the day before. But on the second morning it shifted. Soon gale-force winds had reignited the fire and were blowing it back up the canyon to the east. No lives were lost, thanks to the evacuation, but this time, as it raced up the south side of the creek, it destroyed three houses and then continued to rage up hill into the La Sal Mountains, where it spread over thousands of acres, and where it burned on into the fall. Before it left Pack Creek the fire jumped the road a couple of times, leaving deep black smears on the asphalt that you can still see today. It also left a black charred landscape along the creek itself, the water that gives the place its name now guarded by a leafless black forest of spindly trees. And, after one of those jumps across the road, it found Ken's Quonset hut.

↛ "Everything is gone," he told me when I met up with him a month after the fire. "I lost everything."

All of it, everything in the hut, incinerated, the fire so hot that even the doors of an old woodstove warped. The Quonset hut now looked like a crushed beer can and to walk through it was to walk through a land of ash. The few items that weren't reduced to ash had become outlines of what they had been. An old bicycle now a skeleton of itself.

Of course, Ken and the other Pack Creek residents had not lost *everything*. They had not lost their lives, for instance, like the eighty-five individuals who had perished in the infamous Paradise Fire in California three years before. But one thing they had lost, along with all the property damaged and houses destroyed, was any sense of certainty about a place that had seemed a paradise. And what Ken had lost, his neighbors told me, was the spirit to fight on.

In an effort to restore that spirit, the community rallied. On July 14, a little more than a month after the fire, the residents threw a party to celebrate Jane's birthday, with the secondary purpose of lifting Ken's spirits. The party was held in the Pack Creek lodge, with about forty people attending. The night before, in the same space, the various government and fire authorities had issued their final report to Pack Creek residents. I was there and took thorough notes, but the sentence that most stuck with me was that of a young hydrologist, who assured the crowd that, due to the good soil, the "models suggested" that Pack Creek was not particularly vulnerable to a damaging flash flood. Running counter to this statement was the fact that the phrase "flood insurance" came up several times, making me think that I was back home on the hurricane-threatened coast of North Carolina. Few of these people had flood insurance; it was fire, not water, they feared. Ken's main concern at the meeting was his goats, four of whom had been gone since the fire. He wanted to put together a search party to explore the forest in the mountains above Pack Creek, but the authorities were not allowing anyone up there except the firefighters. "Please look out for them," he beseeched the rescue workers.

The party went well, the wine flowed, and the cake was delicious. I had the honor of meeting and chatting with Clarke Abbey, Edward Abbey's widow. For a good half hour I sat with Ken on the couch in the middle of the lounge. Our topics were age, writing, and uncertainty. I told him about a friend of mine, a brilliant writer, who had recently died much too young. Though Ken was thirty years older than me, I admitted my own anxieties about losing my ability to write as I aged.

"You are going to be writing until you are ninety," he said, patting my leg.

It was then that the rumbling started. Some heard it and immediately went out the glass doors to the patio. They called back in to us and soon we were all outside. What we witnessed next was the most powerful display of rushing water I have ever seen in my life. For perspective, I have watched a half dozen hurricanes up close and stood on the shore as a tidal bore rumbled by in Nova Scotia. This was different. It was a great rush and churn of chocolate-brown water ripping down the creek bed, which it was boring deeper while we watched. The water was like some primitive tool for scribing, and that was what it was doing right in front of us, digging a deeper trough through which it ran. But it was not quite that precise. A kind of liquid violence, it carried tree limbs and moved boulders, charging through with a deafening noise.

It had rained and been cool that morning, a fact that we all celebrated after days of unprecedented heat, and smoke from the distant California fires that blurred the horizon. Afternoon thunderstorms followed the cool morning. But the total rain accumulation was less than an inch—about three-quarters of an inch, according to the locals who kept rain gauges. It didn't matter. Despite what the models had suggested, the fire had denuded enough of the vegetation near the creek to loosen its grip on the land and to help in creating a raging torrent of the sort that no one, not even Ken, had seen before. The water ignored the models. A great brown-red freight train roared down the valley. We all felt awe, but for the homeowners that awe was mixed with fear. Twice in one month their homes had been threatened. First by fire, then by water. A second elemental comeuppance.

The fire had burned the grass within ten feet of the lodge, making the building an island in a sea of ash. The rain now wet the ash as another resident and I walked down through a kind of ashy mud, close to the edge of the rushing creek. From there we could watch in real time as the creek bed was gouged deeper, becoming a canyon ten feet lower than it had been just that morning. The flood rendered the bridge that connected the north and south sides of the community impassable, which left a third of the residents stranded and unable to return to their homes. Only later, when the water went down, would one man ferry the rest across in his truck.

➤ That night it occurred to me that the phrase I had heard so often in my travels—"losing everything"—might have a new meaning. What we are losing may be something much larger than our personal possessions or our homes or even our individual lives. What we may be losing is earth itself, the way it has long been, the way we imagine it will still be, and the way we have lived on it.

It is as if we have broken a contract with the planet. It is hard to see in your own time when a great change occurs. Seismic shifts may feel like regular life just continuing on. But perhaps we really are in a new world. Perhaps we really are in the future that the old books imagined. Perhaps it is happening now.

What if we really can lose everything?

➤ The next morning I got up early and walked down the creek to see the story the water had told. Ken's dog, Boy, barked hello as I passed his house. ("I wanted to skip a step," Ken told me when I asked about the dog's name.) The gentle stream of the day before had returned, but it now ran through a much deeper canyon. Boulders that likely hadn't budged in years had been thrown about, and full trees lay across the water like makeshift bridges. The smell was different. Not quite the burning heat of the day before, but water mixed with ash. A primal, elemental smell.

Flood follows fire. That was the story Pack Creek told in the summer of 2021. It was also the story, writ large, being told all over the West that summer. First, the many fires destroyed the trees and vegetation. Next, the floods swept down unimpeded, carrying the earth with them, gathering up whatever stood in their way.

A western flash flood is something to witness. There are flash floods in the East too, or at least floods that are given the prefix *flash*. But a western flash flood is a different animal. The dry, crumbling land is there for the taking and then the water, all in a rush, comes and takes it. And it takes it fast in one great angry surge, a surge that comes from nowhere and ends just as quickly as it came. By mid-July flash floods were running down burn scars throughout Colorado and Utah, vast torrents gushing down dry, fire-weakened land. One woman was driving down Poudre Canyon outside of Fort Collins and was washed away in her car.

The next morning, I walked over to the sunken bridge, where several residents were shoveling thick mud off its surface. They were making little progress.

"We already had the plague for a year," one resident told me before I left. "Now fires and flood. What next? Locusts?"

PART II.

EMPTY HOUSES

"It's doom alone that counts."

—Bob Dylan

ABANDONED HOMES

A TRIPTYCH

I. RISE

Abandoned houses everywhere.

The first house sits tucked into a cliff of sandstone, the charred ledge above acting as its roof, the one below its floor. It is more a complex than a single home, but the dwelling I am most focused on hides in the shade in the deepest part of the cave. The walls are made from the same material as the cliff they sit tucked into. It is hard not to compare this ancient building to the still-active avian homes that cling to the ceiling above: a small colony of cliff swallows. *Organic.* That's the word. Though to the naked eye the rooms appear empty, the descendants of the people who lived here don't like to think of these as ruins, and you can see why. It doesn't require much imagination to picture living here. You can remember as much as picture it. Like something from a dream that slips away as you wake. But while their spirits may linger, people no longer inhabit this place, not for, what, perhaps seven hundred years, and it's not just a few people who have left. A vast civilization, including one of the first great civilizations in North America, has disappeared from this part of the world. The terms and theories about this disappearance have changed in recent years, along with so much else, but the fact, and the mystery, remains.

About two thousand miles to the east—2,134, to be exact—more abandoned homes, a row of them out on the wet sand of low tide, looking like they are intent on migrating out to sea. Up on stilted long legs, the house furthest out peers down into the water like a great blue heron. But it is a wing-clipped heron. The western wall is torn off and you can see a toilet and a still-made bed, as if, despite the disaster, the owners had left suddenly and might be coming back any minute. They won't. Yellow warning tape flies off the house like the tail of a kite. The house is not where it should be; that is obvious. If you are old enough, something about it might remind you of a scene from a movie, the original *Planet of the Apes*, when the hero, played by Charlton Heston, comes upon the Statue of Liberty, buried up to her torso in sand. It is that incongruous, water lapping the house's legs.

Finally, another empty home. This one much larger than the others.

Unlike the cliff house, it would not require carbon dating to discover when this house, now surrounded with emergency fencing and concertina wire, was built. That would be beginning in the year 1800. Like the cliff dwelling, it sits upon a hill, though the land it looms above is not dry and arid but naturally wet and swampy. Located 359 miles due north of the heron house, it is massive, chalk white, and domed, modelled on the architecture of another lost civilization.

For over two hundred years this house has been the center of activity of the government that makes the laws for the land that the other houses occupy. But now it too is empty. No one home. Back in the desert the house with the cliff swallows assumed a naturally defensive posture, backed as it was into a sandstone wall. This house, too, is playing defense, though in an excessive fashion that suggests impending disaster. Entire city blocks around the house are encircled by emergency fencing topped with razor-sharp wire that glimmers in the sun. Hundreds of cops and soldiers stand guard. They do not smile. Army guards, toting machine guns, stroll behind the fencing. There is no way to get close to the protected house. All it needs is a moat.

These abandoned houses, woven together, tell a story. It is a story that is unlikely to have a happy ending. In fact, it is a story *about* endings. It is also a story, despite the one ancient home, for our time. And finally it is a story, for me at least, about language.

—

Capitol Hill, which shares its name with the building that rests on it, was once the geographic center of Washington, DC. That made it the perfect place to build a structure where the legislative branch of the United States government, the Senate and the House of Representatives, would meet. In its first incarnation, the building lasted just fourteen years before being destroyed in the burning of Washington by the British in the War of 1812, an act of retribution for the burning of both the governor's house and the legislative hall in the Canadian capital of York (which had once been, and would later again be, Toronto). Only after it was rebuilt was its distinctive dome added.

It is March 13, 2021, when I walk up the National Mall toward the abandoned house on the hill. Approaching the building from the west end of the National Mall, I can maintain the illusion of normalcy for a while. At first I don't see all the wire and walls and guns. The sun beams down on my face and people along the mall eat their lunches, throw Frisbees, read, talk, as if the world is still carrying on as it always has. My walk began at the mall's other end, where I read Lincoln's words carved into the walls, and now I've got a good sweat going. In my back pocket is a cheap paperback edition of *Common Sense* by Thomas Paine. I have been fairly obsessed with Paine since seeing the hearings in January. He has helped me to understand how language led this country, not a country yet, through a time of great crisis. I've come to think that Paine might come in handy at this moment. That we can learn something from him.

In the process of reading biographies about the man and

reading his own work, I have so far learned that Thomas Paine was the bestselling writer of his century, the coiner of the name "The United States of America," a founder of the French Revolution as well as the American, and the man who would also eventually give the time he lived in its name: "The Age of Reason." Paine was a self-taught Englishman who hadn't done much of anything before he wrote the pamphlet that set the colonies on fire. That pamphlet, a complex political document told in clear, simple, often-exhilarating prose, was more than any other written work the spark for the American Revolution. Published in late 1775, *Common Sense* sold out in two weeks, and soon there were over 150,000 copies in circulation. The colonies were abuzz with it. The pamphlet was a fierce denunciation not just of monarchs and the monarchy, but of the possibility of future despots and demagogues and the mobs they might inspire. Paine was not alone in his thinking; the framers of the Constitution all wrote defensively, anticipating the actions that might bring down the new and fragile political system known as democracy.

It is a month after the second impeachment trial as I stroll down the Mall. Yesterday I flew up to DC, my first time in an airport since the pandemic struck. The combination of the 9/11 security measures, two decades old now, and the wearing of masks and enforced social distancing, lent a dystopian, nearly paranoid feel to the flight. Suffice it to say that breathing in your own stink from a mask while crowded shoulder to shoulder next to people you have been told you should keep your distance from does not make for a relaxed excursion. After I unpacked in my hotel, I walked out into the empty streets of a city on lockdown. My view of the heavily guarded White House was from almost as far back as the St. John's Church, and all of Lafayette Park was walled off, as it has been since the summer protests. I tried but could not get anywhere close to the Ellipse, the park directly behind the White House where the former president had roused the rabble. And yet seeing this was not really preparation for the many square miles fenced off around the Capitol itself, and the army guards, toting machine guns, strolling behind the fencing.

—

Smells of piñon and sage. Gnarled junipers like arthritic hands. Aridity that sucks moisture out of your every pore.

My regular beat is an environmental one. There is plenty of doom to be found there, and I will get to that soon enough. But I have been thinking lately how inextricably tied the crisis of the land is to the crisis of government. This is nothing new in human history. Wandering the American Southwest in recent years, I can't help but feel I am moving through not just an ancient past, but a possible future. This is not only due to the record heat, historic drought, and rampant fires. It is due to the gone civilization that is always present here.

It is already midday as I approach the ancient home, and to get there I must descend a canyon before ascending. Down the dusty trail marked by twisting junipers with fat blue berries, sage growing from the skull of a rock, prickly pear but also, as a reminder that I am not far from town, an abandoned shopping cart and broken Bud Light bottles on the ground, their shards mixed with the blue of the fallen juniper berries. Further down I enter a world of Gambel oak and cottonwood, and a trickle of reddish water that leads from here to Cottonwood Wash. As I climb I look up at the ancient dwelling, shaded under its ledge—for what, two thousand years?—and I am greedy for that shade. It is not a long hike but I sweat hard in the July sun as I climb. Once there I head to the back of the cave. The coolest spot.

I am not alone. My guide on past trips, my companion on this one, Louis Williams is part Sioux and part Navajo. Louis has come to show me a place where other humans lived for a millennium. An archaeologist recently explained to me why the terms *ruins* and *abandoned* are no longer in favor here. The Native argument is that the spirit of the ancestors still dwells in these places. The argument of the archaeologist I spoke with was similar but more practical. The Ancient Puebloan people, who occupied this landscape beginning over three thousand years ago, were a people on the move, and they would often leave their homes behind. But just as often they, or their ancestors, or other people, came back, sometimes hundreds of years later, adding a new layer of settlement atop the previous one. That is why the term *depopulated* is preferred to *abandoned*, the latter implying the people were gone forever. At first I questioned this but now I question my questioning. Is it so beyond belief that people will once again take up shelter in the safety of these cliffs? If they did it would look, with a long enough perspective, like just another gap between occupations.

Not long ago we would have called the people who lived here Anasazi, a Navajo name. The term is not acceptable today, which makes sense. But "today" is a flicker in time and there is no place where that is more apparent than right here where a thousand years is a dusty second.

Likely whatever I call these people may no longer be acceptable when you read this. But I will say this with some confidence: in this place that we now call Five Kivas, and behind these walls of red stone, humans lived and no longer live.

"Where exactly did they go, David?" Louis asks and his words echo in the cave.

For him at least it is not an entirely academic question.

—

I live far from the desert on the edge of water and land in a region called the American Southeast. The air is different here, thicker, wetter, but we also have our share of empty homes. Not far north of my home is an island called Topsail, which is pronounced *Topsil'*. I travel there from time to time to take stock of the rising sea.

On Topsail I have seen an entire row of houses that looks like it has grown sick of land and begun a single-minded migration seaward. The sheer incongruity of those water houses is startling, and maybe a little thrilling. Walking over the sand toward and then under the buildings you get a sense of something massively out of place. Something awry.

I first traveled to Topsail with Orrin Pilkey, the retired Duke professor whom I have become friends with since landing in the South. The first article I read about Orrin, before I even met him, described how he had thundered his message of sea level rise at the citizens of Topsail. During a town meeting he outlined the three basic choices for any beach community with eroding beaches (which is to say almost *all* beach communities) as being 1) Arming the beaches with seawalls and metal groins or jetties; 2) Dumping more sand on the beach; or, 3) Relocating the houses, that is, retreating. Orrin quickly made it clear what he thought of the first two options, which wasn't much. According to him, seawalls, jetties, and groins didn't work, ultimately destroying beaches, and so-called nourishment was equally ineffective, depending on hard-to-find, high-quality sand.

"I don't believe your beach has a chance of lasting," he said to the homeowners.

Orrin admitted that this was just his opinion, but then couldn't help but add one more thing. "My opinion is better than any mathematical model."

Reading about that meeting was enough to get me to pick up the phone and call Orrin. To my surprise I got right through without any machines interfering or the usual elaborate exchange of messages.

"Orrin Pilkey, world famous geologist," he answered.

Though he didn't know me from a hole in the wall, he was generous with his time. We talked for close to an hour. Before we hung up, he invited me to drive up and continue our conversation at Duke.

Two weeks later he greeted me in his office in the Old Chemistry Building on the Duke campus. Orrin was bearded—a thick Santa Claus–style beard and moustache—and while he was short his stoutness made him appear strong and bearish. As time went by I would stop noticing his lack of height because his personality was so large. He had a deep, gruff voice and he liked to talk. We hit it off immediately.

Over the next fifteen years we would take a series of trips to the Outer Banks, the Jersey Shore and, in the wake of Hurricane Sandy, to New York City. He laughed a lot, despite the whole impending doom thing. And throughout it all he repeated the same message with the consistency and power of a battering ram. *The seas are rising you idiots. Retreat.*

—

Sun glistens off the spiraling concertina wire atop the barricades as I approach the Capitol building.

I am in DC on other business, but what I really hoped to do while here was visit with Jamie Raskin. Today I learned, however, that he, like a lot of us, is working from home, and that the recently ransacked Capitol building is mostly empty. Earlier today he was gracious enough to speak to me on the phone, despite, well despite *everything*. Back during the first week of January, only two months ago, the news came that Jamie's son Tommy had taken his own life. Tommy, who had been named after Thomas Paine, was only twenty-five. He was buried on January 5. The next day the insurrectionists broke into the Capitol. Not much later Jamie was leading the impeachment trial.

I met Jamie when we were college freshmen in 1979. We were acquaintances, not friends. But in 2018 while working on a book about the conservation legacy of Theodore Roosevelt, I sent him an email asking if I could stop by his congressional office in DC. Jamie represented a district in Maryland that bordered Rock Creek Park, where Roosevelt liked to hike and swim while president, and I later learned that Jamie sometimes met his constituents at Rock Creek or nearby Sligo Creek. I would have regretted not choosing to meet outdoors if I had not gotten such a thrill out of being so close to the halls of Congress. I had never been to the Capitol offices before and I had never felt closer to what had been a formerly abstract notion, that of the US government, than while sitting in his office. At the time, with the Democrats in the minority, Jamie felt on the fringes of government and clearly longed for more action. Little did he know.

This morning, when we spoke on the phone, our first subject wasn't the fall of democracy or the prescience of the founders, but sleep.

Or rather the lack thereof.

"I look back on pictures of that week and I look like a zombie," he said.

So how had he stayed so focused and seemingly alert during the hearings?

"Anger was part of it. We were all driven by passion and anger about January 6. We all could have died. Even Lindsey Graham said that before he flipped back to Trump. And there was anger at our colleagues too. That they could put the worship of Trump above their own lives and above the lives of the staff and all the people who work at the Capitol."

By the time of my stroll down the National Mall some are already dismissing the notion that we almost lost our democracy as farfetched. Jamie Raskin is not one of those.

If Donald Trump had succeeded in overturning the election, our downfall would have happened almost exactly the way our founders feared it would. Luckily, because of their fear of this occurring, they wrote a document for the future, a tool that we could use in just such an instance. A kind of time machine fire extinguisher that said "Break in Case of Emergency" on its casing. Short of mentioning Donald Trump by name, they laid out exactly what happened on January 6, 2021.

Our founders, including Thomas Paine, were writers. The documents that guide us are part legalese, part literature. Luckily what a group of revolutionaries thought and then wrote nearly two and a half centuries ago still holds up, though this time around just barely. One distilled result of all this writing was the brilliantly defensive and anticipatory document we call the Constitution. The founders were aware, in ways we are not, of just how fragile a thing democracy is.

"I've got this nervous tic," Jamie said. "I keep going back and reading *The Federalist Papers*. I did that as we were getting ready for the hearings. And it amazed me that right there in *Federalist 1* Hamilton writes about the danger of demagogues."

Jamie had quoted that line from Alexander Hamilton on the first day of the hearings: "The greatest danger to republics and the liberties of the people comes from political opportunists who began as demagogues and end as tyrants, and the people who are encouraged to follow them."

—

We are standing up in the alcove now, among kivas that rise off the rock floor. The kivas form perfect circles in the dust. We face east, toward the canyon we just climbed out of and toward the sun. The people who lived here got to see the sun rise every day but also must have been relieved to have it blocked in the deep heat of desert afternoons. The edge that the kivas sit on juts out sixty feet from the cave in the back, putting anyone standing on its edge at eye level with the tops of the vibrant green cottonwoods that grow down in the wash. Westwater Creek, mostly dry now, sometimes runs down below.

It was and is a fine defensive position. You could and can see anyone approaching from the opposite ridge.

There are active homes below the overhang as well. Clay nests of swallows are plastered to the cave roof, their black entrance holes giving them the look of musical instruments—tiny bagpipes, perhaps. I count twenty-four of these homes. They fit into place perfectly, as do the human dwellings. This is not just a nature writer's sentiment or a New Age sentiment or even a Native one. It is a human one. You can't help but look at this place and say, "It fits." It is *of* the place quite literally, made of the same stone. You could almost believe it grew here if it were not so clearly the work of artists and artisans.

This site is, by local standards, unspectacular. Even a millennium ago, Five Kivas was in the boonies. Today it sits right outside of the town of Blanding, by the juvenile prison, and perhaps a twenty-minute drive from Louis's house. Though it was likely first occupied a couple thousand years ago, it is so close to town that when he was growing up it became a playground for local kids in the summer and was lit up by the pulsing runway lights of the nearby airfield, leading them to name it "The Devil's Heartbeat."

Near the back of the cave is evidence of this more recent occupation. It isn't just rock art you discover if you explore the Southwest but graffiti. We all want to make our mark. Written and even carved into the walls of Five Kivas you can find: *Justin + Dianne* and *Pat Harnon Loves Coco*. And back deep below the fire-charred roof deep in the cave, *Frank*, in an attempt at immortality, has chiseled his name.

When I ask Louis if he has been here before he says: "Nope. Like you, I'm exploring."

This speaks not to a lack of awareness of his neighborhood but to the fact that in the area known as Bears Ears, just to our west, more than 120,000 ancient sites have been discovered. This is a startling thing to a born easterner like me: sometimes you practically stumble across kivas and granaries and whole villages as you wander the desert.

And in a way the very commonness of these places ties in with a point Louis wants to make.

"This land holds important history," he tells me. "Tribal elders tell many stories about this place. But not just history. Some see this as a park to visit, but to me it is a living landscape."

As if to prove his point he picks up a *piñon* nut.

"One of my favorite foods. Delicious and addictive. You can eat it raw or roast it."

I nod. A park you can eat.

"I was fortunate to have grandparents who taught me to appreciate the blessings of Mother Earth. My grandmother sang to the plants while she gathered them and sang to the land while she tilled. She taught me a lot."

Louis walks to the edge of the cliff and points out the features of the place.

"The Abajo mountain waters run down this drainage and eventually into the San Juan Wash. Imagine people being here in, say, 1289, almost two hundred years before Columbus, and listening to the ancestors of these swallows and watching the water below. Sitting up here in this alcove protected from the sun."

The date he throws out is a random one and it is only now, looking back and typing this, that I realize it is smack in the middle of the so-called Great Drought, the megadrought that turned this already dry land to dust. In other words, this land has seen extreme climatic fluctuations before. Here everything hinges on water.

If you were to fly in a plane over the city where I live you would be struck by all the water. Not just the neighboring ocean or the Cape Fear River, which runs past downtown, but the fact that the whole place seems all but saturated, just a few inches above sea level. It isn't hard to find evidence of climate change, and for some we are the poster child of global warming in the United States. But not for me. While that may be true in the near future, at this time we can't claim that title. For the moment that belongs to the desert Southwest, where climate change isn't just present but has been going on for decades.

William deBuys, one of the most incisive observers of the region, writes in *A Great Aridness*: "In apocalyptic visions of global climate change, the North American Southwest makes an easy protagonist, the geographical equivalent of a stalled car on the railroad tracks with a speeding train approaching." There are multiple reasons for this, he tells us, including "catastrophic fires, insect infestations, plant die-off, plant invasion," but the core of it all is the feature that gives his book its title. I'll take the word *aridity* over *aridness* but the point is dry, dry, dry. Bone dry. Dust dry.

When we leave we descend the same trail we climbed. Dust kicks up. The grit of sand and pebbles below our feet. Flies buzz. Swallows chitter. We reach the mostly dry creek. The best years here must have been when the creek ran strong. That would not be this year.

The latest studies say that the current drought has surpassed the Great Drought and that the current climate is the driest and hottest in 1,200 years. Soil moisture deficit, as gauged by tree rings, surpassed even the civilization-ending droughts that were once held up as unsurpassable. Lack of rain is the obvious culprit but less obvious is that the hotter temperatures increase evaporation, leaving the soil and vegetation desiccated. During the twenty-first century, average temperatures have been 1.64 degrees Fahrenheit higher than they were during the half-century before. It has grown worse in the early 2020s with the average temperature rise now 3 degrees higher, and the lack of rainfall still breaking historic records. And of course the obvious differences from megadroughts past: even the most cautious climate scientists, while citing the region's historic variability, now admit that the drought is human-caused.

Historic dates and statistics can sometimes seem as dry as the southwestern soil, but consider this: climate has always determined how people live in this region, and something happened between the megadroughts of the year 800 and the Great Drought. A climatic shift to cooler and wetter conditions, which archaeologist R. E. Burrillo calls "the roaring 1000s," a time of "environmental lushness and plenty not matched before or since in Southwest climate records." This new climate led not just to the growth of crops but to a complex, vibrant civilization. Where

Louis and I walk today was the boonies, but two hundred miles to our southeast, in what is now New Mexico, people gathered to create a place that would serve as the region's capital. The center of that gone world, one of the first great civilizations on this continent and the hub of the desert West for three hundred years, was Chaco Canyon.

That is where I am heading next.

—

In one place not enough water. In another too much.

"The seas are rising," Orrin Pilkey told me the very first time we drove onto Topsail Island. "The time to abandon the coasts is now."

Orrin's predictions for sea level rise were once much higher than those of the Intergovernmental Panel on Climate Change. Over the years the panel has been slowly playing catch-up, though they still lag cautiously behind.

"Seven feet," Orrin said the day we first crossed the bridge to the island. "That's not a prediction, mind you, but a working figure I've now arrived at. If I were in charge of things that is the figure I would use. I would expect the seas to rise seven feet by 2100."

The problem is that that number, floating off in the abstract future, doesn't seem to be able to frighten people into action. Which drives Orrin nuts.

"It's such foolishness. The seas are rising and the storms getting bigger but rather than retreat we draw a line in the sand. The thing is that you can't bargain with or threaten nature. It is no surprise when homes built so close to the water are wiped out. And then when they are rebuilt they are wiped out again. It is a pattern, and while the pattern is just beginning, all the good science says it will continue. So we keep throwing massive amounts of federal money at the problem, money that allows homeowners to rebuild so that their houses can be knocked down again by the next storm."

Topsail Island is, in Orrin's estimation, a particularly egregious example of this foolishness. As we drove along its coast we passed a huge apartment complex teetering over the beach just above the waves, and Orrin grunted as he pointed out the complex's name: *Atlantis*.

—

Time repeats itself but we, caught in it, don't notice. You have to go back to the end to find the beginning.

These days the past bumps into the present, knocking it like a cue ball into the future. Both the building and the name *Capitol* were conscious historical echoes. Even the architectural style is an echo, right there in the name: *neo*classic. Thomas Jefferson, who had as much influence as anyone in designing the building that would hold the government's legislative branch, believed that Roman architecture would set the right tone for the fledgling democracy and that the Capitol's design should emulate "one of the models of antiquity, which have had the approbation of thousands of years." Jefferson was also the one who coined the name "Capitol," which was a word that, until then, did not refer to buildings housing the members of a government.

The word, like so much in the young city, was a reference to an ancient government, one that rose to prominence almost two thousand years before. When Jefferson created this new name he did so instinctively, without much forethought: while looking over the original plans for the city, created by Pierre L'Enfant in 1791, he crossed out the words "Congress House" and wrote in "Capitol." In his *History of the United States Capitol*, William C. Allen writes:

> This seemingly minor clarification was significant, for it spoke volumes of the administration's aspirations for the Capitol and the nation it would serve. Instead of a mere house for Congress, the nation would have a capitol, a place of national purposes, a place with symbolic

roots in the Roman Republic and steeped in its virtues of citizenship and ancient examples of self-government. The word was derived from *capitolium*, literally a city on a hill but more particularly associated with the great Roman temple dedicated to Jupiter Optimus Maximus on the Capitoline Hill.

The building then, like the government it housed, was about the attempt to recover the gone ideals of a dead civilization. Governments grow out of the land, out of climate, but also out of ideas. Of all the biographies and histories I read or skimmed in the month after the impeachment hearings, the one that spoke most directly to my interests was Craig Nelson's *Thomas Paine: Enlightenment, Revolution, and the Birth of Modern Nations*. Nelson sees Paine as an individual but also of his time, a product of the dazzling century before him, a century of social and intellectual change that overturned the verities of the world as it was known.

What had happened during that century? In short, everything. Perhaps its greatest figure, if we can sneak him into an epoch that was mainly eighteenth century, was Isaac Newton, whose discovery of gravity in the late seventeenth had repercussions far beyond the scientific. As Nelson writes of that discovery: "In its wake, a great many moderns would no longer believe in an Old Testament God, who was vengeful, mysterious and inscrutable, but in a Newtonian First Being eminently visible in the glorious benevolence of nature and the astounding beauty of the cosmos." Deism, the notion that God set the world in motion and then let it be, would rise during the next century and become all but the official religion of America's founders. And if God could be questioned, why not kings? What if, rather than divine right, societies could be founded on reason, balance, order?

Meanwhile something not yet called science was on the rise,

as was literacy, and the outrageous notion that all human beings had a right to pursue happiness. In café and pub society, and in the newly founded clubs that were growing like mushrooms, something called magazines, born of a revolution in printing, were everywhere, and popular essayists like Addison and Steele were giving many, even the previously non-elect, new notions about ways to spend their hours on earth, notions that life could be not just less miserable but elevated. A new idea was in the air, a preposterous idea, that people might be rewarded, not for the caste they were born into, but for doing their work well. This idea, which the founders embraced, was called meritocracy. Add to all this the fact that there was an enlivening new drug on the scene that made everything seem all the more inspiring. Among the fitting names not chosen for the time period was the Age of Coffee.

Nelson's book does a fine job of describing the craze for all things ancient, particularly all things Roman, among the educated and self-educated of the eighteenth century, a craze that rose to the level of obsession among the founding fathers. They saw the ancient Romans as models for behavior—*Virtue!* was the watchword—and increasingly for a representative government. General Washington would be hailed as the American Cato (while also later railed against as the American Caesar).

All this was happening in a new language. The first true English novels were appearing. Defoe published *Robinson Crusoe* in 1719, Swift *Gulliver's Travels* in 1726, and Samuel Richardson *Pamela* in 1740. Henry Fielding made his name by mocking the last book on that list with *Shamela* in 1741 but then turned around and wrote something that reads a whole lot like a modern novel in *Joseph Andrews*. In early 1759 two remarkably similar works of short fiction appeared, Voltaire's *Candide* and Samuel Johnson's *Rasselas*. Reading was no longer merely an exercise in deciphering rococo sentences and listening to endless throat clearing, but was full of life, humor, vigor. It would be Thomas Paine's genius to yoke the ordered arguments of the much-admired Roman orators with this new language of life.

As unique as Thomas Paine's *Common Sense* would be, it grew out of this larger literary and intellectual culture. This was a culture that valued complex thought simply and clearly put. The period is known variously as the "long eighteenth century," encompassing the decades before and after that century, or the Age of Reason or, somewhat showily, the Enlightenment. Its writers, from Voltaire to Locke to Edmund Burke to Diderot to the Tory Samuel Johnson, were the literary forefathers of the framers of the Constitution—if not always politically then culturally and literarily—the great document in which the era culminated. It is impossible to generalize too much about a period of literature that included writers as diverse as the Marquis de Sade and Mary Wollstonecraft and Rousseau, but the writing was often brisk, learned, energetic, and reasoned, a language that, like society itself, had begun to see more possibilities in the world, and in government, than previously imagined. They overthrew not just governments but the courtly, roundabout style of writing of the previous era. The new writing was utopian in its search for a better way of being on earth but it was also deeply realistic, aware of human foibles and human avarice. This was the language with which writers like Paine, and Jefferson and Franklin and Hamilton, began to imagine a new country and form of government.

—

Before the fall was the rise.

It is a cold day as I walk amid the great houses with Kialo Winters. Storm clouds bulk up in the west, and big winds and snow are predicted. And of course it is, as ever, dry. Snow, or any precipitation, would be welcome.

For a stretch of close to three hundred years the place we are walking through, Chaco Canyon in New Mexico, was the capital of this dry world.

"They worshipped the elements," Kialo is telling me. "The big wind and the little wind that we breathe in, that plants also breathe. They had found a way and they believed the fate of the world depended on that way. They were searching for the center place, and they believed they had found it here in Chaco. In the center place they would marry the celestial to the landscape."

Kialo Winters is Navajo and Zia Pueblo born, raised on the Navajo Nation reservation east of Chaco Canyon. Louis Williams put me in touch with Kialo, and I am very glad he did. Kialo speaks with the confidence of the teacher he was for over a decade. His knowledge of Chaco runs deep.

He picks up a stick and gets down on one knee and draws a map in the sand. First he makes two crossing lines, signifying the Four Corners. A small rock becomes Chaco, another the western edge of the Grand Canyon, another the Mogollon Rim, another Chimney Rock in Colorado, another Monticello, Utah. He encompasses all these in a great circle.

"One hundred and fifty miles in every direction. The Chaco world was not just the canyon itself. The Chaco world extended out. If an enemy approached the outlying great houses would know."

Since the landscape here is dry to the point of crumbling, it seems at first a most unlikely place to put your center, but to stand in the middle of Pueblo Bonito, the massive hundred-room complex that once rose four stories high in the middle of the desert, is to know you are in the center of things. The great houses, the kivas, the high walls, the sprawl of the place. I use the present tense because it is still here, dusty but standing.

Most people don't know that there was another capital in the land we happen to call the United States. True, they may know about Philadelphia, but few know about Chaco. Fewer still have walked among the ancient buildings and know that Pueblo Bonito, the central building of this capital, is comparable in size to the one in DC.

From AD 950 to around 1250 BC, Chaco Canyon was the center of a web of culture on the Colorado Plateau in the land some now call the Four Corners. Its vast network extended to Mesoamerica, as evinced by the goods that once filled the hundreds of rooms in Pueblo Bonito: turquoise, cocoa beans for chocolate (and caffeine!) and a room just for scarlet macaws. Pueblo Bonito is one of a dozen "great houses" that make up central Chaco, elaborate and meticulously planned structures that contained hundreds of rooms and multiple kivas, deep, circular rooms for ritual and worship and perhaps just plain hanging out. Like another city, Washington, DC, this one was preplanned. The great houses are meticulously aligned with the sun and the moon, the skyline, and each other. Buildings like great sundials and lunar calendars.

"We are not at the center of the universe as a species," Kialo continues. "As organisms we are very thin-skinned. Think of it. Our civilization was just brought to its knees by a tiny bug."

He stresses how important humility, and understanding their place in the larger world of nature and the elements, was to these people. I nod in agreement. And I do agree philosophically and, with regard to the Chacoans, historically.

But I also have to believe something else was at play for a people that created something so obviously monumental, in many senses. There is a touch of arrogance to this center place that might feel a little familiar. It comes out in the origin story that Kialo tells me.

"The ancestors had found a way and the fate of the world depended on humans practicing the way. The ancestors searched for the center place and they found it here. The way would marry the celestial to the landscape and they would do it here. They traveled to many places and said, 'This is not the place to execute the plan.' Finally they arrived here and knew it was the place. Chaco was the place. The first people decided to execute their plan here but the apprentices said, 'We are not sure we can do that here.' What about food? What about *water*? The first people said, 'We'll figure that out later. We need to build *here*.'"

It was, in other words, a shining city on a hill. And like other cultures that deem themselves special, their confidence, bordering on arrogance, led to impressive feats.

—

We hope that our words can save us. Maybe sometimes they can.

Orrin, for instance, has been shooting off sentences like a Gatling gun for decades now. Dozens of books, all with the same theme, the same warning. *The seas are rising.*

It would be tempting to say that Orrin's language is habitually hyperbolic, but time may make a realist of him. He laughs at all the predictions for our future that end after fifty or a hundred years. As if the world will then stop.

He pounds home the same points, hoping someone will listen. He does this relentlessly but with a sense of humor. His is an aggressive and jocular science. He speaks loudly, to paraphrase Thoreau, so that the hard of hearing will understand.

Do they? Do they hear him?

Sometimes.

Language, for Orrin, is how he has fought back against the idiocy. Sometimes he uses it as a bludgeon. Whatever works.

He once told me he had two basic goals as a writer: 1. Fame and glory ("Of course"), and 2. Having a cause.

I replied that I was down with the first goal, but not so much with the second. I added, perhaps a little pompously, that my goal as a writer was to present the messy complexity of what I see, not to offer answers.

Orrin didn't think of writing like that. In fact, he no longer defined himself as a scientist.

"I'm a scientific advocate," he said.

He pointed to James Hansen, who at the time was the head of NASA's Goddard Institute, as a model. Like Hansen, Orrin believed that it was now the scientist's responsibility to speak out politically. Scientists weren't "objective fact machines," nor should they be. Most scientists, out of fear and caution and careerism, refuse to offer their opinions about the facts they uncover. But who better to offer opinions than those who spend their lives studying a thing?

Over the last thirty years scientists feared making bold statements about the climate crisis: make a big statement and you were sure to be met with scurrilous attacks. Orrin's role, it seemed to me, was to do just that, be bold, without fear, and part of that role was acting as a sort of human shield for other coastal scientists. As he learned way back when he published his first book, he didn't mind criticism. He could take it. Laugh it off. Even *like* it a little bit. He knew there were scientists doing more relevant, and perhaps better, work at the moment. But he was a bit like the coach who deflects criticism from his players.

Scientists are vital of course, and Orrin was a good one, but in this world where scientific truths are often ignored, someone needs to help make them heard. That's Orrin's new job. A kind of human bullhorn. It's a job that requires a quick tongue and a thick skin. The toughness to sit in a town meeting where everyone is readying their torches, and the wit to appear on *The Tonight Show* if called upon.

What he could add, but didn't, is that he had also become, despite a belief in uncertainty, a professional prophet.

Orrin is comfortable with his role as doomsayer in a way I am not. Over the last few years, almost despite myself, my subject as a writer has become the end of the world. I'll admit that sometimes I find it a troubling beat. As a young man, I wanted to write novels and never imagined myself chronicling the story of the rising seas or the overheating earth. Maybe we'll all eventually be forced out of our comfort zones, though many will cling to what was, an older vision.

I remember a strange moment, a decade old. I was walking out of a class at the southern university where I teach and eavesdropped on two young students having a laugh over the apocalypse. As I walked by the common area that served as a lounge for our students, this was what I heard:

"She really believes that the world's going to end, that, like, the warming's going to come and the water's going to rise and drown us all or something."

"Yeah, like *that's* gonna happen."

I was sure the latter comment was accompanied by an eye roll, and as I passed the two young women I indulged in the internal equivalent of the same. *The fools! The nonbelievers!*

I experienced a moment of happy superiority, followed by the spasm of frustration common to those who believe their great and obvious truths are being ignored, a feeling no doubt experienced by some of my fundamentalist or evangelical students when they regarded me, their godless Yankee professor.

The warming! I would snicker over that phrase later with my wife. But at the same time I thought I understood the young women, at least a little. There was in what they said, if not a grain of truth, then a grain of what is a common attitude toward any prediction of massive global change. This is an attitude, not just of skepticism, but of outright disbelief.

This is the way the world is, we think, *the world we know, and this is the way the world will stay.*

To say that most of us are a sort of climate change skeptic is not to say that most of us doubt the work of science. What we are perhaps truly doubtful of is the ability of our species to predict the future.

True skepticism, after all, is the bread and butter of both Orrin's profession and mine. Scientist and essayist.

It's a question I've been wrestling with: How to reconcile apocalypticism with skepticism?

—

There is a sort of craving that you don't know exists until it is sated, one where you are not quite aware of what you needed.

I hadn't understood, until that moment back in the second impeachment hearings in February, how hungry I was for sentences. Balanced sentences, packed sentences, sentences spoken by someone who wasn't ashamed of having read a book every now and then, sentences, though it didn't occur to me immediately, like those written by the founders of this country.

It might have started when he quoted Voltaire. Or maybe Thomas Paine. Not the famous "summer soldier" speech that he concluded with, but an earlier mention, one that I haven't yet found in the congressional transcripts, but that I'm sure is there. It wasn't so much that he was referencing Enlightenment thinkers—that's a fairly pat move for politicians—but *how* he did it. He didn't appear to be reading the quotations off a teleprompter, but rather seemed to be summoning them from memory for the occasion, for the *moment*, and did so casually, offhandedly, and with a deep familiarity with the words, just like the college professor he had been for a quarter century. It felt like what it was: an intelligent person talking to us like we were intelligent people.

And here is what I felt, watching Jamie Raskin at home on my TV: a profound sense of relief. Not that I thought that our fractured country would suddenly be healed or that the Republican senators would actually vote their consciences. No, I wasn't delusional; that is not what left me feeling relieved. It was the words themselves. The sentences. I realized how hungry I had been during the Trump years, starved really, for balanced sentences and distilled thought. Not grunts and groans and bullying repetition and ad hominem attacks but well-crafted phrases and seamless references to, and quoting of, thinkers from a time before ours, thinkers who had created this country and who it turned out had handed down the exact tools we needed at the moment.

In that instant I saw the impeachment hearings in a new way. Not just as a response to an attack on the Capitol, or even an attack on the Constitution, the very document that was meant to guide the January 6 confirmation of the Electoral College vote. As Jamie Raskin and his team made clear, it had also been an attack on the ideals, and thoughts, of the founding fathers, who feared nothing more than a mob led by a demagogue, a new self-proclaimed king. But it was also, perhaps less obviously, an attack on language itself. On the words and sentences that the founders set down almost two and a half centuries ago, words and sentences that reflected the Enlightenment values of those who wrote them, words and sentences that were meant to light the way of those of us who came after.

From the beginning the language of liberty was the language of struggle. It was never going to be easy to keep the flame alive. Fittingly, Jamie Raskin concluded his closing speech at the impeachment hearings by quoting Thomas Paine. After amending the first line at the suggestion of Nancy Pelosi—"These are the times that try men *and women's* souls"—he launched into the sentences that Paine had written in *The Crisis*, when the revolution looked hopeless.

"The summer soldier and the sunshine patriot will shrink at this moment from the service of their cause and their country; but everyone who stands with us now will win the love and the favor and affection of every man and every woman for all time. Tyranny, like hell, is not easily conquered; but we have this saving consolation: the more difficult the struggle, the more glorious in the end will be our victory."

The hearings sparked my own mini-obsession with Thomas Paine. I learned that Paine has been the most overlooked of founding fathers, but at the moment of the publication of *Common Sense* there was no more celebrated man in the colonies. Everyone was suddenly talking about the pamphlet, even its detractors like John Adams, and it is generally considered the nudge that was needed for the United States declaring itself an independent country. Thomas Jefferson said it most simply: "History is to ascribe the American Revolution to Thomas

Paine." The introductory paragraph alone, penned a few months after the original had already sold thousands of copies and been reprinted, announces not just the birth of a country but of a writer with voice, a voice that is like the one that Philip Roth described in *The Ghost Writer*, the kind "that begins at about the back of the knees and reaches well above the head."

Paine writes: "Perhaps the sentiments contained in the following pages, are not yet sufficiently fashionable to procure them general favor; a long habit of not thinking a thing wrong, gives it a superficial appearance of being right, and raises at first a formidable outcry in defence of custom. But the tumult soon subsides. Time makes more converts than reason."

There are scholars who look at the Declaration of Independence as simply a more formal retelling of *Common Sense*. Some of Paine's biographers argue that he merely caught the wave of revolution when he published his famous pamphlet in December of 1775. Others argue he created it. Whatever the case, it helped move the idea of war with England from something almost incomprehensible to something inevitable. Reading it now you can, if you extend some historical empathy, still see it, still *feel* it. The writing is modern and direct, particularly when it comes to ridiculing the idea of hereditary monarchy. He made the whole notion of kings, then considered divine and the embodiment of nations, seem ridiculous, dethroning them with vigorous prose: "Men who look upon themselves born to reign, and others to obey, soon grow insolent; selected from the rest of mankind their minds are easily poisoned by importance; and the world they act in differs so materially from the world at large, that they have but little opportunity of knowing its true interests, and when they succeed to the government are frequently the most ignorant and unfit of any throughout the dominions." Kings as ignorant! Imagine the shock of his excited readers. And notice the combination that was his trademark: the balanced phrases and plain language. (Paine's original title for the pamphlet was *Plain Truth*.)

Common Sense is one of the preeminent examples of written language changing history. It is a great distillation of Enlightenment thought that focuses on the taking down of kings, the lifting of the common people and their right to pursue liberty and happiness, all mixed with some very practical suggestions about creating a government and the need to declare independence. Many of these Enlightenment ideals had been building for a century, but they had never been put this way before. The philosopher Alfred North Whitehead once criticized what he called "inert ideas." Paine's ideas were anything but inert. They were put forth with panache, grit, energy. They were alive. They fired people up. They changed things.

Is such a thing still possible?

—

It wasn't just drought that did in Chaco. Societal disintegration and disease helped too, and the people's proclivity toward movement and migration. But as old theories replace new, as they always will, it is best not to throw out the drought with the bathwater.

To stand inside Pueblo Bonito is to see civilization in the past tense. It doesn't take much imagination, however, to bring it alive again and picture yourself back at the center of things. All of the rooms that spread out before you are massively and monumentally impressive. If you explore cliff dwellings in the Southwest, one of the pleasures is the way they hide from you, tucked into crevices. This place hides from no one and is tucked into nothing. It is not small or simple, hidden in the rocks, but something that stands out. Something that was meant to stand out.

One of the several mysteries of Chaco is the question of what the hell was going on here. My friend Craig Childs, whose groundbreaking book *House of Rain* tackles this and other questions, writes: "The evidence gathered from a century of digging and mapping can support almost any speculation thrown at Chaco Canyon: religious center, military center, government center, ceremonial center—the list is extensive." He goes on to say that he has heard some call it "an ancient Las Vegas, an isolated strip of grandiose architecture in an ill-watered desert where people came from all directions to participate in flashy ceremonies and where they left all their wealth before heading home." But if it was Vegas, it was also DC and New York. (Our history books claim that the first four-story buildings on this continent were in New York City, then Chicago. Our history books are wrong.) All roads led to Chaco, and these were real and well-tended roads, spreading out in all directions and all the way out to places as far away as Five Kivas. Outliers likely paid homage, and taxes, to Chaco.

Standing in the bone-dry basin of Chaco another mystery comes to mind. Why here? Though a wash runs through it the wash too is dry, as is the rest of the place to the point of desiccation, and the buildings sit out in the open in a landscape where little but sage and rabbitbrush grow. Part of the answer is that it hasn't always looked like this. This is important. Like the United States itself, Chaco had the good luck of early abundance, its development growing out of a quirk of climate that tree rings reveal, that string of good days and months and years that led to an abundance of crops in 950, which in turn led to a surplus and the creation of the complex of kivas and great houses. Back then the dry wash I am staring down at ran wet. It was this stroke of good elemental fortune that allowed for what came next. For the more than 250 years Chaco ruled.

During that golden period, this truly was the center place.

II. FALL

As well as being where our laws come from and a neoclassic building constructed in 1800, the Capitol building is where Jamie Raskin goes to work.

Jamie could have been forgiven if he played hooky on the morning of January 6. On any other morning that would have been a laughable notion. This was a man who, for the better part of four decades, talked about constitutional law and politics the way other men talk about their beloved sports teams, a man who was weaned on politics by his activist father, Marcus, a member of the Kennedy administration and the founder of the country's most influential progressive think tank, the Institute for Policy Studies. Jamie followed his father's political lead, fighting against the Reagan administration's interference in Central America as an activist Harvard student in the early eighties, becoming an editor of the *Harvard Law Review* at Harvard Law School, teaching as a constitutional law professor at American University, serving as a state rep in Maryland for two terms, and, finally, winning a seat as the Democratic representative of Maryland's Eighth Congressional District in 2016.

On the same day that Jamie became a congressman, another man, who knew somewhat less about the Constitution, was also elected. The day of Donald Trump's inauguration as president of the United States was one of the few days in Jamie's life when he actually did play hooky. Instead of driving into Washington he led a hike of his new Eighth District constituents down a winding trail along Rock Creek in Rock Creek Park, taking, as he put it, "the idea of Democrats in the wilderness to a whole new level."

That had been a protest, a form of political expression deeply familiar to Jamie. The morning of January 6, 2021, was different. That morning, under the powers of the Constitution he had so long studied and loved, Jamie would cast one of the votes certifying the results of the Electoral College, results that would end the reign of a man whom he regarded as a stain on American democracy. In other words, he would use the document he revered and had spent his life studying to evict the man who he believed had spat on that document.

So of course he would go. Of course he would cast his vote.

The reason that it might have even occurred to him not to was that this day, which would prove one of the most tragic and powerful and overwhelming of his life, directly followed a day that was even more tragic and powerful and overwhelming. The day before, he had buried his beloved Tommy. Tommy, his beautiful twenty-five-year-old son, who had followed him to Harvard Law School, and into activism and politics, and who seemed to light up the world and inspire everyone he encountered. Tommy, who loved his parents and sisters so ferociously and who always made them laugh and who had vowed to fight for the poor and for the rights of animals and who to his father was not just a wonderful son but a political visionary who "was ten thousand years ahead of his time," but who also, suffering from depression, "felt every bit of pain and injustice in the world." On New Year's Eve, not a week before the historic Electoral College vote, Tommy had taken his own life. He left a note that said: "Please forgive me. My illness won today. Please look after each other, the animals, and the global poor for me. All my love, Tommy."

For Jamie the pain was piercing at times, fog-like at others. He couldn't sleep. People sometimes call suicide a selfish act, but Tommy's sister Tabitha said that the pain of depression must have been so great that it even outweighed the pain Tommy knew he would cause his family, and that if he was being selfish, "it was the one selfish act of his life."

Tommy had not just been named after Thomas Paine. He embodied the same utopian spirit, the same vision of a better world, as his namesake, and, as it happened, the same proclivity for depression. Tommy the activist might have described himself just the way his father did in notes for his twenty-fifth college reunion, as an "unreconstructed Enlightenment liberal."

It was Jamie's idea that Tabitha would also come into the Capitol on January 6. The night before, she had said: "Don't go to work tomorrow."

"I've got to go," Jamie replied. "Why don't you come?" In a

speech a month later on the first day of the impeachment trial, Jamie Raskin would explain why he had invited Tabitha and Hank Kronick, his son-in-law, who was married to Tabitha's older sister. Hank had never been to the Capitol. Jamie's reasons for inviting them were high-minded: they would join him to "witness this historic event, the peaceful transfer of power in America." But something simpler must have also been at work. Tabitha must have wanted to support her devastated father. She and Hank wanted to be with him on the day after he had put his only son in the ground.

That morning Tabitha wondered aloud about the rumors of violence by Trump supporters.

"Will it be safe?" she asked.

"Of course it will be safe," Jamie replied. "This is the Capitol."

—

Over the years, Orrin and I began to travel farther afield. Everywhere we went we would see what I first saw on Topsail repeated on other islands, most prominently on the Outer Banks. Everywhere massive development coupled with rising seas and dangerous storms.

Once in the town of Nags Head we pulled over at a spot Orrin knew, a spot that showed just how little respect the ocean had for all the building that was going on.

A dozen or so houses stood out on the flat, low-tide sand. If not for the fact they were on stilts, some of the rooms would have been in the water at high tide. The signs of desperate self-defense were evident in piled sandbags, but the bags were obviously long past usefulness, shining green with algae and half underwater. Useless electrical wires and pink insulation hung limp from the undersides of the houses. The houses themselves, stranded out on the low-tide beach, distanced from their usual surroundings of roads and neighboring homes and telephone wires, had the look of sci-fi space stations floating far away from earth. Stairs ran down off the houses and hung in the air, hovering above the water, and "Condemned: Do Not Enter" signs shone orange in the windows. It was a truly wild sight, no less wild for the fact that the structures were manmade.

We walked below them, staring up at old rusted doors that opened out to nothing and stairs that dangled in the air. Many of the abandoned houses were lopsided, or tilting backward as if popping wheelies, and some looked ready to take their final stagger and lurch into the water.

"When they fall there will be rubbish and insulation all over the beach," Orrin said. "Nobody cleans it. The insurance companies won't go near them until they are declared officially ruined."

Not fifty feet back from the ocean we came upon a concrete septic tank half buried in the sand, a great square sepulchral tomb of shit. Orrin kicked it. Then he pointed back at the houses and explained that they were the third generation of homes to have ambled off into the sea. Of course new homes were still being built. The next group of water marchers.

—

Beginnings are times of excitement. Endings are something different.

If you want to contemplate endings, and to see things from a distance in dry air, this is your place. If it is hard to imagine time on a larger scale, Chaco helps.

Still, we don't like to think about endings, about doom, about things falling apart, and I get that. "What are you avoiding?" any good psychologist will ask. What we, the people, are avoiding is where we are so obviously hurtling toward. Those signs of our institutions breaking down just as the physical world seems to be turning on us. And there are plenty of good reasons to avoid this. I, for one, am not by nature an apocalyptic thinker, and would prefer to consider the score of last night's Celtics game or what I'm having for lunch. But even I can see it.

Let's say for a minute that you wanted to contemplate the fact that *it* could really happen. That our civilization, along with our country's government, could fall apart, just as every civilization has fallen apart throughout history. If so, you could do worse than coming to this corner of the desert Southwest. Here you can see and think about a civilization that has gone away due to climate, yes, but also due to a breakdown of institutions, infighting, and war. I said a civilization, not a people. The people migrated and evolved and are still here in the form of their descendants. That is why the Chacoans are currently referred to as Ancestral Puebloans, and the culture continues in some form. But the civilization that was is gone.

Standing in a small stone room that has withstood centuries of sun, wind, and even occasional rain, you can think about time. As you study the red stone that was first piled up centuries ago, a raven breaks the silence. It feels like there is nobody else but us around for miles. Here you can see time entombed.

I try to work my way back, to imagine the place when it was the center of its world.

As with the young United States, *ideals* prevailed at Chaco. But as with the United States, these ideals did not apply to all.

Wealth inequity, Kialo Winters explains to me, is nothing new. As we walk along the beautiful sandstone walls of Una Vida, the earliest of the great houses at Chaco, he describes an archaeological dig here in the early twentieth century where the scientists hired on local Navajo workers to do the digging.

"What happened at Chaco has been in our stories, our oral tradition, over the millennium," he says. "Science never asked locals what happened here."

But one day during lunch break sometime in the early 20th century, one curious archeologist did.

"Lunch break that day lasted a lot longer," Kialo says.

One of the things that the workers already knew, but that would take a while for science to figure out, was that a caste system prevailed at Chaco, with the equivalent of lords and ladies ruling over the equivalent of peasants or perhaps slaves. How did scientists confirm this? The skeletons inside the rooms of the great houses were taller and buried with many relics, those outside smaller and buried without.

The stories got there first. Kialo tells me the story of the Gambler, named Naahwillbiihi ("winner of the people") or Noqoilpi ("he who wins men at play"). The Gambler came to Chaco from the south and soon was besting the locals at many games of chance, including dice and games of skill, like footraces. Once he had taken everything else, he would always make one final bet and take their souls. What did he do once he had their souls? Perhaps force them to labor as slaves in the creation of the monumental architecture that abounds at Chaco.

The story continues that the Gambler himself was finally beaten by a Navajo shaman, who exiled him to the south.

We are told we are now in the midst of a crisis of climate, but it is more than that. It isn't hard to connect this to a crisis of government or to imagine how the two crises, working in concert, might play out. All you have to do is look backward.

While our Gambler rests up down in Mar-a-Lago, we await what comes next.

—

Jamie's son-in-law Hank would have a front row seat at the chaos. That morning Hank was excited to walk the marble hallways of a place he had only read about or seen on TV. The fifteen-minute drive from the hotel where they had stayed took about an hour due to the barricades and hordes of protesters, who, Hank noted, did not wear masks. Even the most cynical of us can experience a thrill at being in the citadel of democracy, and despite the threat of the mob, Hank was caught up in it all. He, Jamie, and Tabitha headed up to the fifth-floor office of Steny Hoyer, the House majority leader. If anything had been bipartisan in this most partisan of times it was the outpouring of sympathy over Tommy's death, and colleagues from both sides of the aisle dropped by to offer their condolences. Jamie finally had to close the door to Hoyer's office to finish a speech on unity he was about to give on the House floor. Tabitha, looking over a hard copy, crossed out a few lines, saying they were "too divisive."

When it was time for Jamie to head down to the House floor for the vote, he hugged Tabitha and Hank goodbye. They stayed behind with Julie Tagen, Jamie's chief of staff, who was growing increasingly nervous as she glanced out the window at the gathering protestors. Tabitha and Hank joined her and as they watched the mob grow larger, climbing up the steps and then up the inauguration scaffolding, they debated whether or not to just go home. Hank would later write: "After a week of grieving, we were not well equipped to deal with fear or uncertainty. We were constantly reminded that we were in 'the safest building in the country' and that it would be more dangerous to go outside. After all, the vice president was right down the hall from us. This gave us peace of mind; the idea of members of the mob getting inside was unfathomable."

When they headed down to the gallery to watch Jamie's speech, they were escorted by Jason, a plainclothes security detail. In the chamber Jamie stood to speak and was greeted by a standing ovation from the entire House. After the speech, Tabitha, Hank, and Julie headed back to the majority leader's office. Once they were inside, Jason said to them: "I'm going to lock you in here." He didn't tell them why.

But both the window and the TV revealed the same scene. Hordes of angry protesters were now yelling, "Stop the steal!" and trying to break in to the Capitol. Rioters were literally climbing the walls. Right before the TV went dark, the power having been cut, Julie, Tabitha, and Hank saw that the building had been breached, and moments later they heard the intruders rumbling down the hallways inside the building. Hank remembers going into survival mode:

Do we try to escape? I checked the bulletproof glass windows and realized they were sealed shut. We were stuck with no way out.

Do we fight back? I glanced over at a fireplace poker, briefly envisioning myself as the hero taking on the mob head-on before realizing this would be a foolish fight against the hundreds of angry men and women right outside our door.

Do we hide? This seemed like our only option. Praying that if we stayed hidden and silent, the intruders would open the door and leave.

—

"You're about to officially become a climate refugee," Orrin said when I called to tell him I was evacuating with my family during Hurricane Florence. Hurricanes were the one time the name of our town appeared in the national news, and for the previous week the only place names on the weather maps had been Bermuda and Wilmington, my town. It was coming straight for us.

While I was evacuated during Florence up in Durham, I visited Orrin. It just so happened it was his birthday.

"Not every day you turn eighty-five," I said, going by what Wikipedia was telling me.

"Hey, eighty-four, eighty-four!"

"Okay."

"One resource the world is not running out of is old people," he said.

I picked him up at the retirement home. It's like college, he told me. You've got the popular kids. The loners. The cliques.

I took Orrin out for a birthday lunch at an Italian place down the street. The sausage sandwiches were a revelation.

"Hey, I've only come here for dinner," he said. "Now I'm going to start coming for lunch."

Over lunch we talked about the damage done by Florence.

"The Outer Banks seem to have survived," he said. "For now."

I asked him about the term "climate refugee," which was still relatively new back then.

"It refers to those who leave and don't come back," he said. "Like the 250,000 people who left New Orleans during and just after Katrina. When Hurricane Harvey hit Houston, 20,000 of those Louisiana refugees were still there. Homeless again."

I thought how western fires have done this even more effectively than southern storms. Entire towns erased from the map.

Orrin predicted that in the coming years two million climate refugees will come from the Mississippi Delta and four to six million will come from south Florida. Also a couple hundred thousand from the outer and inner banks of North Carolina.

"These are the three areas that will flood most quickly," he said, "at least here in the United States."

Orrin was encouraged by some examples of increased awareness since our last trip up the coast together. In Norfolk, where flooding is so frequent, they had begun to list the tides in the church programs. In New Jersey, the state was purchasing at-risk coastal homes, demolishing them, and making the land public property. In Charleston and Manhattan, they were building seawalls.

"I thought you hated seawalls," I said.

"I hate the ones that destroy beaches. It's okay for Charleston and New York and Boston to build seawalls because there's no beach there to worry about. So you don't have to worry about the question of which is more valuable, buildings or beaches."

I mentioned Miami, which seemed to have dodged a bullet again this year.

He shrugged.

"Not much you can do there. Miami is doomed."

On the way back to his place we stopped at the Eno River and watched the rising water. All around the state the waters were still rising and my route home was impassable. In the end, flooding would kill more people than the storm itself.

Orrin was living the last years of his life inside the predictions he had made.

—

The government rests on the land. Literally of course, but also figuratively. It has always been so. Since our beginnings as a species.

Here is how the archaeologist R. E. Burrillo describes Chaco:

> A massive rapid civilization development. Suddenly you've got big buildings, big communities, roads that extend for miles, a trade network that extends to Mesoamerica, and all that in a relatively short period of time. Which correlates with at least one environmental phenomenon, which is the extremely rainy period around AD 1000.

Climate gives and climate takes. Chaco was the closest thing to the Rome of the Southwest for hundreds of years, and while drought was long thought to be the chief culprit in this Rome's fall, other contenders have emerged. Like Rome, this capital had vast cultural and religious influence on the outlying settlements, and like the famous peace of the Roman Empire—*Pax Romana*—the period when Chaco thrived was a relatively stable and peaceful one. It didn't last. R. E. Burrillo writes in his book, *Behind the Bears Ears*, that "the so-called Pax Chaco disappeared in a hurry. Archaeological evidence from this period indicates an extreme uptick of interpersonal violence throughout the region that's almost undoubtedly associated with a major breakdown of social influence or control."

—

Tabitha was the one who leaned the chair against the door's handle, though it didn't seem like much of a defense as the forces of chaos gathered outside. She huddled with Hank under House Speaker Hoyer's desk. They held hands. The noises in the hall grew louder. They began to think they might die. Hank texted "I love you" to his parents. The door shook as someone tried to open it. That was the worst moment in a night of worst moments. Julie, Jamie's chief of staff who was trapped with them, turned fierce and protective. She grabbed a fire poker and said, "I'm not letting these motherfuckers get away with this."

Meanwhile, back on the House floor, Jamie heard a noise he would never forget. It was a blasting sound, a pounding on the House chamber doors like "a battering ram." Jamie was frightened for his own safety but couldn't stop thinking of Tabitha and Hank. Having lost his son, would he lose his daughter and son-in-law? The night before, disconsolate after the funeral, Jamie had slumped in a chair and told Hank: "You're the only son I have now."

Not far away a police officer was being pummeled with an American flag.

—

As might be apparent by now, I like to jump around. Not just as a writer but as a reader. Lately, for instance, I have been dividing my time between books on Chacoan history, Thomas Paine, and the architecture of the Capitol. It's always a strange pleasure when diverse things intersect. This morning, reading Mr. Burrillo's book, I learned that in Chaco societal breakdown was likely marked by viscious infighting in the great houses, with the capital eventually moving sixty miles north to Aztec.

Then I read this:

> A fair-to-decent comparison would be when the British burned the United States Capitol building in Washington during the second year of the War of 1812 after which a new capitol was built to the north in Philadelphia under the direction of a nervously and aggressively galvanized leadership that took a more iron-fisted approach to foreign and domestic policy than its

swampy predecessor had. This didn't really happen of course—although it almost did.

Some of the intersections are even more unlikely.

Many of the rooms at Chaco were eventually used as tombs for the leaders, filled with thousands of so-called grave goods, including precious turquoise, macaw feathers, and pottery. In one small room, thought to rival the tombs of Egypt in terms of pure wealth, fourteen people were buried vertically along with tens of thousands of precious relics. It turns out that this game of intergenerational piggyback included seven men and seven women and spanned 330 years.

The heroes of the Age of Reason looked down on such primitive doings. Though not always. Take the Capitol building itself. Yes, the architecture is neoclassic and the thinking behind the building that of the rationalism of the Enlightenment. But something darker lurks below. A crypt built for our first president lies directly beneath the Capitol Rotunda, which forty Doric columns support. This is where the remains of Washington would have decayed, had his son not scotched those plans and insisted his father be buried back home at Mount Vernon.

—

Is it farfetched to think that Thomas Paine was already on Jamie Raskin's mind not just when he composed his arguments for the impeachment trial but while he was under assault in the Capitol building on January 6?

Maybe. Maybe his thoughts were focused solely on survival. Survival not just for himself but for Tabitha and Hank. But he was trapped by the mob not for minutes but for hours, and there was plenty of time for his mind to roam. And if it had roamed, it would have roamed to thoughts of Tommy. It had only been six days at that point. Tommy had taken his life the Thursday before, New Year's Eve, and it was just Wednesday now. Only yesterday they had buried him.

For forty-five minutes Tabitha, Hank, and Julie stayed hidden as the noises outside the door grew louder, then subsided. Then came a knock on the door. Someone identified himself as being from the Capitol police. They had to decide if it was ruse, if these were really members of the mob trying to lure them out. After ten minutes they cautiously opened the door and found three officers standing there. The officers escorted them through the halls of the Capitol to the safety of a ballroom turned shelter where Jamie and others were waiting. They raced toward each other and hugged and then Tabitha said the words to Jamie that he would quote a month later at the impeachment hearings: "I never want to come back to the Capitol." That was the quote, but what she really called it was "*this place*."

"That day I learned how personal democracy is," Jamie would say in the same speech. "And how personal the loss of democracy is, too."

—

One mystery of Chaco was why all the wood in the abandoned buildings was charred. Kialo explains that when the people left they "closed up" Chaco, ritualistically caving in the roofs of kivas and burning the great houses. There are other explanations, but I like that one.

During our time together, I learn that Kialo and I share a peculiar obsession. We are both deeply interested in how long various civilizations have lasted.

He points out that Rome was the sole capital of its empire from 27 BC to AD 286, which is 313 years.

Chaco had a similar run from 900 to 1150 AD. Two hundred fifty years.

Washington, DC, became the capital of the United States in 1800. Two hundred twenty-one years as of this writing.

"Maybe our number's up," I say when he lists these dates.

He laughs.

—

Washington, DC, the city, was famously low and swampy and may be so again. The working figure for the Intergovernmental Panel on Climate Change for sea level rise by 2100 is one to three feet. Many, including Orrin Pilkey, believe that number is far too conservative, not taking into account the possible melting of the Greenland or Antarctic ice sheets. Many more sober-minded scientists have argued that the melting of the ice sheet would be unlikely. Until this year.

When I first started traveling with Orrin I fact-checked his sea level prediction of seven feet by the end of the century by calling James Hansen at NASA. Hansen assured me that seven feet was a "good working number" as we prepared for what was to come. And that was over a decade ago, before anyone really believed Antarctica was truly vulnerable.

"The glaciers are the key," Hansen said.

Orrin is in the habit of rattling off scary numbers. If the West Antarctic Ice Sheet melts we can expect an 11-foot rise, if the Greenland ice cap goes, 13 feet, and if all of Antarctica were to melt, the global seas would rise 170 feet.

If this last were to happen you could kiss the White House, at an elevation of 59 feet, and the Pentagon, at 71, goodbye. Also, all major East Coast cities.

The Capitol building, at an elevation of 288 feet, would manage to keep its head above water.

It would sit up on its hill like an island.

—

Unbelievably, even after his reunion with his daughter, the day was just beginning for Jamie Raskin. Members of Congress, vowing to not let the insurrectionists stop the vote, reconvened at eight that night. Seven senators, including Ted Cruz, still objected to the certification, which did not pass until three that morning. As a constitutional scholar, Jamie knew better than almost anyone else in that room how precariously close we had come that day to losing American democracy. There was always the possibility that the president would declare martial law in the chaos he had created.

The final electoral vote took place on the one-week anniversary of Tommy's death.

Jamie did not get home that night until four a.m. but he was up again at eight. He had already been drafting the legislation to invoke the twenty-fifth amendment and now he set to work on drafting the article of impeachment. The next week Nancy Pelosi would ask him to lead the House's impeachment team, and he could not say no. Though he was still barely sleeping he felt Tommy with him, "his presence, his strengths, his beliefs."

It was as if Jamie had been built for exactly this occasion; it was what he had trained for, and even by the end of that first day it must have occurred to him, despite the chaos, the sadness, the tragedy, that what he had witnessed was not just an assault on the Capitol but on the Constitution itself, the very document that he had devoted his life to studying.

For the time being that document, written 233 years ago, has proven sturdy enough. But for how long?

—

For anyone paying attention, the past has begun to feel very present and has maybe even begun to feel a little like the future. Here, in this land of the great houses and cliff dwellings, aridity is still the theme played year round except for the brief bursts during monsoon season that flood this dry land. Wallace Stegner wrote: "Aridity, more than anything else, gives the western landscape its character." In the desert Southwest the rising of the sun has long been both worshipped and regarded with a well-earned nervousness, its ascension marking both the beginning and end of something. The oppressive heat that other parts of the country has felt recently has long been felt here. And if this extreme place becomes even more extreme? When temperatures hit 120? 130? Can human beings really continue to live here? They have left before.

R. E. Burrillo writes of the end of Chaco: "That decline co-oc-curred with a climatic plunge, during which local and far-flung residents lost faith in the established sociopolitical system."

He adds: "It wasn't pleasant."

Here at Chaco the past suggests a possible future. Not that we will listen. Stuck in our virtual present we lack a historic imag-ination. It can't happen, we insist. But it has happened already. In this place, and all over the world. Why would we consider ourselves exempt from ending? Because of all our modern stuff? *Please*. We think of time as an arrow not a circle, but take a look around and you'll see us circling back.

I am in no way predicting how we the people will come to ruin. I am predicting nothing.

But without pushing it too hard you can see hints here in Chaco. And parallels. As I walk over the dusty ground through Pueblo Bonito, a structure that was once four stories high, the word *monumental* comes to mind. I think again: this is not the work of a humble people. In fact, I am getting a little *Masters of the Universe* vibe. This is the work of people who wanted to be known, who wanted to be remembered.

Rina Swentzell, who grew up in Santa Clara Pueblo and has a master's degree in architecture and a PhD in American Studies from the University of New Mexico, has written an essay that gets at just this. The essay, "A Pueblo Woman's Perspective on Chaco Canyon," is collected in the book *In Search of Chaco*, edited by David Grant Noble. In it Swentzell writes of her first encounter with Chaco at the age of twenty-eight:

> Even then my response to the canyon was that some sensibility other than my Pueblo ancestors had worked on the Chaco great houses. There were the familiar ele-ments such as *nansipu* (the symbolic opening into the underworlds), kivas, plazas, and earth materials, but they were overlain by a strictness and precision of design

that was unfamiliar, not just to me but to other sites of the Southwest. It was clear that the purpose of these great villages was not to restate their oneness with the earth but to show the power and specialness of humans.

For me, they represented a desire to control human and natural resources. They were not about the Pueblo belief in the capability of everyone, including children, to participate in daily activities, such that the process is more important than the end product. The Chaco great houses projected a different sensibility. The finished product was very important. Skill and specialization were needed to do the fine stonework and lay the sharp-edged walls. I concluded that the structures had been built in the prime of life with a vision of something beyond daily life and the present moment. They were men who embraced a social-political religious hierarchy and envisioned control and power over place, resources, and people.

The preferred theory these days for the so-called disappearance of Chaco is contained in the current name for these people, the Ancestral Puebloans. The idea is that these people did not disappear but migrated and are still among us, part of today's Indigenous population. But Swentzell suggests not mere continuance, but evolution and even rebellion against the ideals of Chaco.

Here is how R. E. Burrillo put it when I asked him about what happened:

> Societies don't simply collapse. They change. They alter. They might not take on the same shape forever. So you can point at them and say yes they are still here. And at the same time they are gone. Chaco is a good example of that. A lot of the same folks I've been leaning

on say that much of what we see in "pueblo" cultures, Hopi in particular, seems like not so much a continuation of Chaco as a reaction against it. The arrogance of Chaco, the class system. You see the same thing among the Maya. The great city-states went away and the Maya essentially said "fuck it."

In other words, they had achieved the heights and made that achievement their way of being, and those who followed said, "Um, excuse me, there is another way. Yours is not the only way to be on earth. It might even not be, sacrilegious as it is to say, the best way."

This, too, could be more than a little applicable to coming times. What happens in the wake of an arrogant overweening civilization? People do not simply disappear. But they do, in all likelihood, change. As they always have.

The people still living here hold another secret. On these Native lands, adjacent to the largest reservation both in the country and the world, they understand how it feels to come out on the other side of an apocalypse. It won't do to romanticize a people who have endured so much, so I won't. But I will say that, for what it's worth, they know a little about something that all of our children, or our children's children, might have to learn.

That is, they know what it is like to live *after*.

It is easy to fall for the trap of thinking the fall will lead to a de-peopled world, like Will Smith having New York City to himself in I Am Legend. But Italy isn't empty because Rome fell thousands of years ago, and villagers still have the blood of ancient senators in their veins. So too, modern Puebloan and Zuni and Hopi and Navajo people. They are still here, but have been through their own brand of apocalypse, one that began two hundred years after the fall of Chaco. War, displacement, disease, genocide. A people bludgeoned by fate. If we are worried about what is coming, we only need to look and see what has been.

—

These days Orrin's language has begun to seem less hyperbolic.

In the fall of 2021 the United States Department of Defense, Homeland Security, and the National Security Council issued separate reports on climate and its effect on this country's security. Taken together, the picture these reports painted was grim. Heightened military tension. Massive dislocation of tens of millions of people. Fights over scarce resources. Water. Food. Potential economic collapse. In the course of less than a year the country has experienced eighteen individual climate disasters—fires, storms, floods—that cost over a billion dollars each.

The year 2021 saw historic fires in the West, a massive hurricane that ripped through New Orleans and flooded New Jersey and New York, the earliest tropical storm, the latest first snow on Colorado's front range, melting chunks of glacier in Antarctica, cyclones in December in the Midwest, sixty-degree days in Alaska in January.

Trying to retain hope and optimism is one thing. Seeing clearly is another.

If you have not been directly affected, you know someone who has.

And we are just getting warmed up.

—

It would be nice if Thomas Paine's story had ended with victory. But having been a founder and a witness of the French Revolution, and having waited in a jail cell condemned to death for treason by the French in 1793, Paine would come to understand the ways in which the word *freedom* could be usurped by the mob. His question is still our question: How to make people understand that that sort of brute thinking is, at root, antidemocratic? That freedom is not a slogan but something that is hard-won and that the winning of it comes through tough-minded thinking and discipline. Something worked toward until one gets it right or close to right. The founders got it right enough that the documents they created still work today.

As Jamie Raskin put it at the hearings: "Donald Trump doesn't know a lot about the founders but they sure knew about him."

The documents the founders handed down to us, the Declaration of Independence and the Constitution, are imperfect in ways that are obvious now, starting with the fact that the "all" who were created equal were a limited group. Tom Paine, one of the few founders who never owned a slave, fought aggressively against that so-called institution all his life. "A slave republic" is what Jamie called the United States in his closing remarks. But the basic thinking, and the writing, of the founders is solid and self-corrective enough to still guide us, and can possibly still light our way. Those words were the ones that Frederick Douglass used to argue against slavery and the ones that Lincoln built on in the Gettysburg Address, and the ones Martin Luther King used as the beginnings of his cries for freedom. Flawed, yes, but prescient in ways we, ingrown and blinded by our moment, can barely imagine. Prescient enough to anticipate a Trump and prescient enough to allow for their own evolution. As Jamie Raskin said in closing:

> However flawed the founders were as men in their times, they inscribed in the Declaration of Independence and the Constitution all the beautiful principles that we needed to open America up to successive waves of political struggle and constitutional change and transformation in the country so we really would become something much more like Lincoln's beautiful vision of government of the people, by the people and for the people—the world's greatest multiracial, multireligious, multiethnic constitutional democracy, the envy of the world.

Tom Paine said the United States was "an asylum for humanity" where people would come. Think about the Preamble, those first three words pregnant with such meaning: "We the People" and then all of the purposes of our government put into that one action-packed sentence: "We the People, in order to form a more perfect union, establish justice, ensure domestic tranquility, provide for the common defense, promote the general welfare and preserve to ourselves and our posterity, the blessings of liberty do hereby ordain and establish this Constitution..."

What Jamie was suggesting was yet another evolution of our original thought. This was reflected in his own multiracial team of men and women, the House managers who came together to fight against just the sort of threat, just the sort of tyranny, that the founders had warned us about.

I end by circling back to language itself. At the key crisis points in our country's history, at its birth and when the country split in two, words have come to the rescue. It is language and those who wield language that have saved us. From Paine to Hamilton to Jefferson to Lincoln to King to Obama, our leaders have been writers. This makes sense. The way we write is the way we think. The way we think influences how we act. Therefore it is a matter of immense importance to put things as well, and clearly, as one possibly can. If the founders had not done so we would not be living in a democracy.

Whether language can sustain us going forward is another question.

—

Three houses, all abandoned.

A house walking out to sea. A house tucked in a cliff. A house where the laws of our land are made.

Three houses, three languages, three stories.

Three tragedies—though, amid the rubble, you can also, if you are so inclined, find three traditions. A scientific tradition that insists we see the empirical truths of cause and effect. A Native tradition that sees beyond the merely human and treats the land as sacred. A shopworn but still-living political tradition of celebrating Enlightenment values, truths of liberty and freedom, not caricatures of them.

We will need them all, fully braided, in the times ahead.

But that is too hopeful a sentence for an elegy.

The truth is we will only know where we are going when we get there. Predictions are mere guesses. A man did not walk out of his stone home in Chaco, stretch his arms above his head, yawn, and think: "This civilization is about to end." He thought about his crops, his family, about gambling on a friendly game of bone dice on the Chacoan equivalent of next Thursday night.

The language of beginnings is intoxication. That of endings is naturally more sober, even somber. We try to rekindle what we once had. Sometimes we succeed. Mostly we don't.

"The prophecy for the Navajo is that the world will end in fire," Kialo tells me.

Yes, and the prophecy, if you live in my neck of the woods, is that the world will end in flooding.

And so we return to three empty houses.

In one the water laps at the base of the stilts that hold it aloft, while the tide continues to rise.

In the next desiccation spreads outward, like the rays of a blazing summer sun, into the ever-drying region it was once capital of.

And the final house? Its fate is the least certain of the three. We have seen it shaken by just the sort of attack its founders feared. Will its foundation prove sound enough to withstand the storms ahead?

It is not for us, trapped as we are in time, to know.

SAFE PLACES

My younger brother doesn't recognize me when I walk into the Staples where he works in Hyannis on Cape Cod.

"Can I help you?" he asks as I stand in front of his register.

I just stare at him for a second. It is May 2021, my first time back on the Cape since the pandemic.

"My name is David Gessner," I say, hoping this will clear things up.

"Oh my God, you look so old," he says.

This stings a little, of course, but I know that part of it has to do with the mask I wear, as if I've come not to say hello to my sibling but to hold him up. The mask scrunches my habitually scrunched eyes upward, accentuating the wrinkles below my eyes until I'm sure I look a little like the Emperor on *Star Wars*. Though I still believe in my heart that I am a young man, I am not.

Cape Cod, that stomping ground of my youth and twenties and later thirties, is also a well-known home for retirees, which is the age I'm closing in on. Who knows? Maybe Nina and I will move back here and settle, spending our last years in the place we once thought we would call home. Though the fact that now even relatively modest homes on Cape Cod are out of our price range argues against this.

My brother has, in one way at least, fulfilled my dream, having lived on the Cape for close to a decade now. However, his is no pastoral fantasy by the shore. He has had a hard life, marked by battles with his own mind, reduced expectations, and, more recently, cancer. Twice I have visited in recent years to be with him during operations.

My brother's search for home is less metaphoric than mine. He lives in a single room where his landlord periodically threatens to kick him out and then changes her mind. His roommates have had their own battles and more than one has done time in jail. He has one good friend, over in the town of Dennis, and that makes me happy. Earlier in his life, in Austin, Texas, where he had been getting a graduate degree, and then again in Boston, he spent some time homeless on the streets. Ironically, he also worked in the housing authority in Boston, winning a prize for his work.

He longs for a new place to live, but good luck. There is no room in the inn called Cape Cod. No homes, no rooms, anywhere, he is told. This on a peninsula where thousands of houses sit empty for ten, sometimes eleven months a year. If you want to see a physical embodiment of income disparity, this is your place. Whole neighborhoods of large homes wait empty through the winter for their summer inhabitants to return. Then the population more than doubles and the large empty houses fill.

The pandemic has made it worse, as it has in so many rural places. The locals have to compete with affluent owners of second or third homes who can now Zoom to work. In this seller's market, landlords are cashing in and leaving renters houseless. With fewer homes for sale, prices have shot up, a 38 percent increase in median house prices in the last year. Now the average house on Cape Cod goes for $630,000.

In contrast, the great Cape nature writer John Hay paid twenty-five dollars an acre for the one hundred acres of so-called worthless woodlot he bought in the 1940s before heading off to war. John died back in 2011, and that woodlot is now conservation land in the town of Brewster.

I have come back to Cape Cod in part to work on a novel about John's life, and have spent many hours this week tramping around the land that was once his. It is late May, and as I walk the paths he made by years of hiking, the white stars and mayflowers glisten, rhododendron flowers burst, titmice call to each other. I am pleased that John's land is now a commons.

John concentrated much of his writing, and thinking, on the search for home, and our perpetual failure to find it. He believed this country, and the world, had become detached from the land, "being always on the move." The fact that they had no place to call home also left people morally adrift.

"The worst thing we've been doing over the past years is to forget about localities," he told me during an interview for a book I wrote about him. "You forget about localities—individual places—and it's much harder to find out where you live. When I first moved here, I knew an old Cape Codder who said this place was getting filled up with people who didn't know where they were. And when you are displaced you start to think everything is money—money is the only purpose for so many people now and they just accept this mindlessly."

These days John's ideas might be casually discounted because of his background. A child of affluence, his grandfather and namesake served as both Abraham Lincoln's secretary and, decades later, Theodore Roosevelt's secretary of state, and it is true that John's concerns about what he called the "vast epidemic of homelessness" were substantially more theoretical than even mine, let alone my brother's. But he was brave in his own right, choosing an artist's life that his family found questionable, and committing to his worthless woodlot for over sixty years. I saw him as a model once, back when I thought there might be a way for me to spend my life wedging down into this sandy land that juts out like a hook into the Atlantic. It would have been a good life. But it wasn't the life I ended up living.

—

We do not feel safe where we live, my wife and I.

Maybe you know the feeling.

And maybe you, like us, look toward other places with dreamy eyes.

Twenty-five years ago I published an essay called "A Polygamist of Place."

I stand by it.

In calling myself a geographic polygamist I meant that, while many of the writers I admired—from John Hay in Brewster to Wendell Berry in Kentucky to Henry David Thoreau in Concord—were so rooted in one place that they seemed married to it, I was never able to settle down in one place, having multiple places I loved and never being sure which one to commit to.

Moves have marked my life. My dark secret as a nature writer is that I didn't grow up on the shore or in the woods but in Worcester, Massachusetts. During my thirtieth year I moved back to Worcester. The official nickname of the city is "the heart of the Commonwealth," but some of us prefer the unofficial one: Wormtown. Wormtown seemed an appropriate name that year, since I learned that I had testicular cancer and thoughts of being worm food were never far from my mind during my operation and month of radiation. "I don't know what's worse, cancer or Worcester," I wrote in my journal, and I wasn't entirely joking.

Boulder, Colorado, is where I moved after Worcester, where I grew healthy walking up the trails into the mountains, and then, once I discovered that it was against the law to be out of shape in Boulder County, *running* up them. Colorado, to borrow a trope from John Denver, was where I was reborn, trying on a new identity as a westerner. I thought I would remain in that state forever.

But Boulder is another failed home.

It is early June of 2021 when I fly here from Cape Cod, or rather from Boston to Denver, and make my way here. Hadley and Nina will catch up in a couple of days, but on my first morning back I climb up to the base of the Flatirons and turn back east to watch the sun rise over the plains.

This is done as a tribute not just to the red ball that is rising, but to a good friend. It has been quite a year for death and one of my hardest was Reg Saner's. Reg, my former teacher, died on April 19, during one of the season's last snowstorms. Reg made it an almost daily practice to hike up to the Flatirons and watch the day begin. One of his books was called *The Dawn Collector*, and a dawn collector he was.

One dawn that Reg, Nina, and I shared occurred on September 11, 2001. It was still dark when he picked us up in his truck at the cottage we were renting in Chautauqua. Our destination was the Indian Peaks Wilderness, and had we not stopped on the way at a convenience store (for tampons for Nina, she has allowed me to say), we would not have heard the strange news that a plane had apparently flown into a building in New York. For nine hours, with only a vague sense of what had happened, we hiked to the top of the world. For me, this was a moment when nature and disaster became one. We climbed to the Arapaho Glacier and then up to South Arapaho Peak, which at 13,397 feet was ten times higher than the burning buildings. Reg was seventy-three years old when he made that nine-mile hike with a three-thousand-foot elevation gain. For the whole day we were both in brilliant light and in the dark. That night, at a pizza place back in Boulder, we finally saw the footage of the planes hitting the towers.

Reg lived close enough to the Flatirons that a short walk would bring him to a spot where he could collect his dawns, staring back toward the treeless plains and Denver to watch the sun emerge. I decide to emulate him during my first two days in town. Dawn, he wrote, "is the rock my church is built on, and such soul as I have is stored sunlight. Through them I feel the depth and range of this world compared to our superficial sense of it."

This morning, as I watch the red edge of the burning star sneak over the horizon, I couldn't agree more. The sun lights up the land behind me. At this hour and in this golden light, the Flatirons feel like a mythic landscape. Thrilled to be here, I begin to scheme and plan ways to get back, ways to live here, almost forgetting that I am here now.

When I come back down the hill I also come back to my senses. Or maybe, more accurately, I leave my senses up on the hill behind me and return to what we call good sense. If Cape Cod is unaffordable, then this town, where my wife and I met, is impregnable. The median price for a home here makes Cape Cod look like a bargain. During the pandemic, Boulder home prices increased by 50 percent. It now costs, on average, 1.55 million dollars to live where you can walk up the hill to see the sun rise over Denver. "Our houses made more than we did," Reg used to say, referring to the appreciation of his home, bought in the sixties, and those of other longtime professors at the University of Colorado.

The sun continues to rise as I hike down. In retrospect, my years in the West seem idyllic, though as I get older I'm inclined to argue that "in retrospect" is the only time when idylls exist. Seven years after coming West, I moved from Colorado back to Cape Cod. My first book was about the Cape and when it came out I felt it wouldn't do to be living two thousand miles away, though soon after moving back I began to write a book about how much I loved Colorado. My geographic uncertainty was showing. Still, I took some of the West back east with me. I had learned some things during my western years, and read many books, including most of those that Wallace Stegner had written. Stegner was a master of seeing the arid western landscape both intimately and through a large lens, and his vision was informed by long acquaintance and a deep study of climate, geography, and history. I wondered whether I might be able to apply the same principles to the sandy coastal climate of the land that I had loved as a child. Nina and I were newly married, and we began to imagine spending our lives on Cape Cod. Though we both still longed for the West, if I was ever going to marry a place here it was. The tides had gotten inside us and we loved the way that the bird migrations swept through during fall and the way the return of the seals from northern waters signaled the beginning of another raw off-season.

Then Nina got pregnant. We were both writers, not just unrooted but in debt, and we had somewhat foolishly made writing books, not making money, our main priority in the world. I was now in my forties and our daughter was on the way. After six years on Cape Cod, a place we could have lived forever, we briefly moved to Boston so I could teach, and that spring, a few weeks before Hadley was born, I accepted a full-time job teaching in Wilmington, North Carolina, a town I had never heard of.

The new place, on the hurricane coast of North Carolina, overwhelmed us at first with its heat and strangeness. We were parents of a newborn child whom we loved, to our surprise, more than we had ever loved anything. So we put a good face on it, and before I knew it our exile to the South was offering up surprising possibilities. And birds. They helped immensely. Egrets and pelicans and ospreys and oystercatchers and clapper rails and skimmers and herons of all stripes and, not to be forgotten, strange, scimitar-billed ibises.

When we first moved south we rented something called a unit, not a house, and on the entire nearly treeless island of Wrightsville Beach there was barely a single home where you couldn't stick a broomstick out the window and touch your neighbor. The southerners thought it paradise, but we weren't so sure. For one thing there was the heat: it slammed us, stunned us, slathered us in sweat. Everything wilted, and we were part of everything. It was the kind of heat that made you want to lie down and give up, to throw up your arms in surrender. The heat even seemed to stun the pelicans, those strange new birds, as they flapped lazily and deeply, rowing along as if through something viscous.

"It's like Viet Nam," we told friends back north, until we started to feel guilty about it, since neither of us had been to Viet Nam. But when I mentioned this to another friend, a Viet Nam vet, he reassured me.

"I spent a summer in the Carolinas," he said. "It's worse than Viet Nam."

I tried to maintain an attitude of relaxation when we first moved to the island, but at times I felt as if I'd lost my balance. The heat had begun to make me act in questionable ways. Each morning when I lifted Hadley into our un-air-conditioned Honda Civic, I felt as if I was sliding her into a kiln. This led me to a place I never thought I'd go. While I was an animal known as a nature writer, I was on the verge of doing something unspeakable and preposterous: buying an SUV. It was a small SUV, but still. I had raged against these abominations most of my life, but now I found that the cars we really wanted to buy, the cool green enviro station wagons, were out of our price range. The main thing, I told myself in a near panic, was to get my daughter out of the kiln.

I was ready to buy it, to pull the trigger, when I was saved, at least temporarily, by an unexpected source. The Toyota guy himself called with some news. Our credit report had come back and our loan had been rejected.

I asked why.

"You have weak stability," he told me, reading from the report.

Yes, of course.

I nodded and considered the poetry of his words.

But there were already other moments, moments when— despite my wobbly confidence—I had glimpses that the southern coast might not be such a bad place to live. With summer ending the parking lots began to empty: there were fewer beach-walkers and more pelicans. I remember one morning carrying Hadley down the beach in the little chest papoose I often wore in those days, and watching two immature, first-year pelicans as they flew right down over the waves, belly to belly with their shadows. It was exhilarating the way they lifted up together and sank down again, rollercoastering, their wings nicking the crest of the waves. Eight more adult birds skimmed right through the valley between the waves, by the surfers, sweeping upward before plopping onto the water.

That calmed me, but the new world I found was still an uncertain one. We now lived in a place with a Confederate heritage where storms could lay the place flat and where only a few inches of sea level rise meant water would swallow the land. Climate change quickly grew more personal. A hurricane hit after our first month, and the girls evacuated while I stayed, ready to go down with the ship. Human plans and goals might seem primary to human beings, but I was learning that in a blink, and by the whim of wind, my plans could be rendered not just secondary or tertiary, but entirely unimportant. In the end I felt less like Henry David Thoreau than Robinson Crusoe, tossed up on an unfamiliar southern shore, certain of nothing. I began to study the storms and the rising seas. Nina and I dreamed of our past homes even as we adapted to our new one.

Gradually I saw a possible way to exist in this new world. A way to be in uncertainties. It was in the South that I redefined myself. Had I stayed in the North I might have also stayed within a literary tradition of extolling one's home place above all others. This is a tradition I greatly admire, but one that, on Cape Cod at least, was worn out and well-trod and not just by John Hay but by generations of male writers from Harvard starting with Thoreau himself. Wendell Berry, speaking of his beloved Kentucky, once wrote that one of the challenges of writing about that state was that it was so unwritten, devoid of tradition. The Cape I loved had the opposite problem: it was overwritten, and you couldn't find a walk that hadn't already inspired an essay or poem. So for all its heat and other challenges, living in the South was freeing. What's more, I didn't really *leave* the West or the Northeast, but traveled there frequently, and there was an energy in that moving back and forth. It was around this time that I was approached by an environmental magazine about writing regularly for them, and I often found myself, and my writing, on the road. I more fully embraced my role as a traveler, not a settler.

—

Here is the thing.

Boulder was never just Boulder. It was a jumping-off place to that mythic land of my dreams: the West. It was where I launched my trips to the mountains of Colorado, the deserts and canyons of Utah, Arizona, and New Mexico. These were the trips where I felt most alive. And where I still do.

I will get to my own environmental hypocrisy soon enough, but let me say for now that part of the thrill of the West was *driving* to places. Having grown up in Massachusetts with Route 28 and the Pike, I only knew driving as the interlude between when you left and when you got there. Here was something new. Drives like massive hikes through deep canyons and over mountain passes filled with golden shimmering aspen leaves and down into lands of orange rocks the size of skyscrapers that twisted and turned like no rocks I had ever seen, and always something new, and usually something startling, around the corner. And then to decide to pull into one of these places on the spur of the moment and stroll right in because, after all, they belonged to *me*. Yes, there were public lands in the East but here there were millions of acres that I was part owner of just waiting to be explored. Southern Utah in particular pulled like a magnet. Who knew such a startling place existed? Back in Worcester no one had told me.

In retrospect my environmental hypocrisy is obvious. The dinosaur fossils that were fueling my trips were also burning up the world. In North Carolina I like to make fun of the Weather

Channel reporters who, while standing out in the storm, tell others that no one should be standing out in the storm. My own hypocrisy was using fossil fuel to get to places to write about the evils of fossil fuels. I have no defense. In the past, when wrestling with these issues, I have quoted my friend, the environmentalist Dan Driscoll, who was responsible for the greening of the corridors along the Charles River. Dan said: "We are all hypocrites. But we need more hypocrites who fight."

What Dan meant, I think, is that too many of us, noting our own eco-flaws, throw up our hands and say, "What's the point?" But if only those with spotless environmental records fight for change, then we will have very few fighters.

I am aware that this does not get me off the hook. As I have said before, on the hook is where I am and where I belong.

—

Though I have traveled through the West for decades now, I am not prepared for what I find when I drop off Nina and Hadley at the airport and push off from Boulder. They are headed back to North Carolina and I am headed out on the road. The stated reason for my journeys is a book tour, and at many of the stores I visit, mine is the first in-person talk since the pandemic, people coming cautiously out of their caves. But fairly quickly I see that my real purpose is not to shill for an old book but to report for a new one. And what I find, as I travel through Colorado, Wyoming, and Utah, is startling. I think I have seen it all, but I haven't.

Smoke from the fires is everywhere, the place a blur. Never has it been so dry and hot. I do mean *never*, or at least never since the great drought that altered civilizations seven hundred years ago. Once, while stopping by the side of the road, I pick up a dry clump of dirt and crush it in my fingers. It rains down like dust, and provides a pretty good stand-in for the region itself.

On the one hand this is nothing new. When I began to explore the West, thirty years ago, I fairly quickly discovered the large-scale serpents in that billion-acre garden. The West wasn't just massive but massively fragile. An arid and breakable land, prone to disaster. The experience of exploring the greater West was always Janus-faced. Show me a beautiful place and I will show you the threats to it. I had fought my share of environmental battles back East, but in this new place there was a different feel. In the West the threats were sometimes as spectacular as the landscapes. You could emerge from a canyon and find a uranium mine. In a place that looked like it would exist as is forever you could hike by tailings of an element with a half-life of 4.5 billion years. It wasn't as if this destroyed the joy of exploring the vast landscape but it certainly complicated it. I would climb to the top of a river overlook in a national monument one day and drink beers with workers in a fracking town the next. Over the years, mine was a long absorbing education in the realities of the West, an education that stripped the place of much illusion, but never entirely of its romance.

This year is different. The whole region is burning up. It is a conclusion you can't help but reach if you are traveling through the West in the summer of 2021. It isn't just the fires and floods, but the extreme heat. It's like your kid with a bad fever only your kid is the land you love. Meanwhile smoke is everywhere. The mountains and mesas and deserts, clouded by the smoke from the many fires, look like developing Polaroids of themselves. The whole region smells charred. The day I cross the Colorado River my car thermostat registers 106 and that same day Death Valley hits 130.

It's like the future has arrived. *Thunk.* While I have observed and written about climate change for many years, this is different. And something beyond the climate itself has changed. It is as if we are, all of us together, in unison, crossing a kind of mental barrier. When I talk to people, I've never heard so many nevers. "Never have we had so many fires." "Never has it been so hot." "Never have we had so many juniper trees die." You hear it on the news, you hear it from friends and family, and then later in the summer, with perfect timing, we will hear it from the world's leading climate scientists when the 1,600-page Intergovernmental Panel on Climate Change (IPCC) report on climate lands with a thud.

Traditionally, people look at these reports as warnings. This new report will be a warning, too, but it doesn't need to point to the future to scare us. It simply says, "This is what is." The conclusion that the world's top scientists reach is the same many of us have: after years of debating about climate change, we are now living inside it. In fact, the first two sentences of the report will state: "It is unequivocal that human influence has warmed the atmosphere, ocean and land. Widespread and rapid changes in the atmosphere, ocean, cryosphere and biosphere have occurred." For those familiar with the earliest IPCC reports, this is a new language. Not just the use of the word *unequivocal* but the use of the past tense: *have* occurred, not *will*.

Then, on the same day I cross the Colorado River, my wife calls from back home in North Carolina to tell me that the first tropical storm of the season is about to hit. Carolina is not immune to the nevers. "We have never had so many storms," my wife says when I talk to her. It is true. When we first moved to North Carolina, hurricanes came occasionally and did not feel as threatening. Back then my wife and daughter would evacuate and I would stay behind and ride them out, and there was a kind of thrill to the experience. That had been the early days. Now the thrill is gone. Everything changed with Hurricane Florence in 2018. What was once occasional feels annual, at the very least, as every summer we ready ourselves. Ready ourselves for what? For the possibility that our lives will be washed away.

East. West. Fire. Water. One place too dry, the other too wet. And both getting more the way they are. Reverting to their essences.

—

Part of loving places is feeling pain as those places are diminished.

Most of the professors who taught writing at the University of Colorado when I attended seemed curiously disconnected from the place where they lived. Reg Saner was an exception. Another exception was Linda Hogan, an essayist and poet and member of the Chickasaw Nation. In Linda's work she often looks beyond the human to the world of trees and animals, and it was in her class that I first began to write about birds.

Birds have always provided a connective tissue between the places that I love. From the pelicans and skimmers of the southern coast, to the magpies and towhees of Boulder to the canyon wrens and ravens of the desert to the ospreys of Cape Cod. Like me, most birds are migrants, and it has been painful and personal to watch their numbers drop in places I know well.

One can barely imagine how these losses feel to one who is truly native to these places.

Linda Hogan writes about loss on both a personal and deeper historic level:

> The long histories of invasion and conquest have changed cultures, created the extinction of peoples and languages, cleared the forests of medicines that would be significant now if not for this breaking and the many losses....Now we face other challenges faced by all people as polar ice melts and Native villages are flooded and moved, as mines and fracking use up the aquifers. Forests are cleared and the water disappears from cloud forests....When water leaves, it has to move to other lands, creating floods, hurricanes, and tsunamis that endanger and kill the inhabitants of those places. And where the forests once existed are now areas of surprising heat and drought. Life is disappearing through the hole we have created that grows over our shared world.

What we don't know about this earth is still large. Old knowledge that we need to learn anew is leaving us. In this changed world, the regard for life is now often missing and our work is in learning to bring it back.

—

After my trip I circled back to Boulder. Coming down into the valley of my former home has always provided a moment of lift for me due to the sight of the distinctive foothills called Flatirons. Not this time. When I drive in it is as if the Flatirons have grown shy. They are nowhere to be seen, hiding behind a veil of smoke.

The noxious air and smoke obscure not just the Flatirons but the Rockies behind them. For a while people have been saying that Boulder is ruined: too rich, too crowded, too white. Despite this, I have clung to a romantic vision of this place where I got healthy again. But now the air itself is poison. It is getting harder to find any places you can feel safe.

Usually leaving this place fills me with regret. But later in the summer when I board the plane in Denver I will be ready.

Like so many of us, I just want to go home.

THE ROAD TO PARADISE

I need to see where all the smoke is coming from.

After a few days of rest in Boulder I head west again.

Many have pointed out that we are in the midst of another megadrought, exceeding even the Great Drought of 1200. The recent droughts and fires also bear a resemblance to another more recent event. As with the Dustbowl, when the clouds of dust blew all the way to DC, announcing themselves to our lawmakers, the fruits of this summer's disasters, in the form of smoke from the many fires, have blown back to Minnesota and Massachusetts.

My experience on this trip, as with all my western trips in recent years, is a double or perhaps triple-sided one. A deep delight in the places I find myself and in many of the people I meet, an anger at the despoiling of these places, and a deeper sadness about the climatic future. That's a lot to roil around in your head, and the miracle is how often, despite everything, delight wins out.

It is still a beautiful world.

The food is good, too. On my way farther west I stop to have dinner at the Hell's Backbone Grill and Farm. My host is Blake Spalding, the outgoing co-owner of the restaurant and the nearby farm where the restaurant gets much of its produce. Hell's Backbone is located in the tiny town of Boulder, Utah, on the edge of the great redrock wilderness of Grand Staircase-Escalante National Monument, but is considered by many to be

Utah's finest restaurant. The restaurant, and Blake, were profiled in *The New Yorker* in 2018, and Hell's Backbone continues to draw food-lovers and celebrities from all over the world. Blake has also become an at-first-reluctant activist after Donald Trump decided to dramatically reduce the Grand Staircase-Escalante National Monument, which her restaurant all but borders. This created a clash not just with Trump, but with many of the locals in town. With charm and humor and passion, Blake navigates the split between the different groups of town residents, a split that perfectly reflects our national one.

The next morning I decide to stop and hike along the Escalante River in the Grand Staircase-Escalante National Monument. The drive into Grand Staircase from the north, along a knife's blade of stone with thousand-foot drops on either side, is not for the faint of heart. I don't like heights and I have to drive slow, slow, slow, not looking down, taking deep breaths to still my panic. My breathing changes when I get to the river, however. I get out to walk by the water for a while. And then I just keep walking. This is a spontaneous, unplanned hike, the best kind, and I run into no one. Freedom, the real kind, is still possible in this compromised world, if you can only keep your brain quiet for a while. I find that I can as I wander along the purling Escalante through the sagebrush while staring up at the great battlements of stone. Above me swooping swallows feast on insects. I take a break from the apocalypse to birdwatch under the giant gnarled cottonwoods. More swallows, and woodpeckers, towhees, rufous hummingbirds, and a chickadee-sized bird I can't identify. They are all that matters for the moment.

A swim suggests itself and I strip down to just my sneakers (it's rocky) and climb into the lazy river. If it's not paradise, it ain't bad. Still wet, I devour a turkey sandwich I've brought along while sitting in the sand beside the river under the shade of giant mushrooms of pink-tinted rock. Prints of deer and coyote mark the sand. Perfect little hollowed-out alcoves provide the shelves, like an apothecary desk, on which I put my water bottle, binoculars, and journal. I stare up at sheer sandstone walls that rise five hundred feet above. The sun blisters down but here in the shade it is actually quite cool. I make a half-hearted attempt at a nap.

This is what I need. For a couple hours I can believe there are Edens left on earth, and both the experience itself and the fact that such a place exists buoys me.

But. The inevitable *but*.

Back on the road the feeling fades. Heading toward the town of Escalante, the reality of this burning summer returns. I sense it in my nose first, then my eyes. Down by the river I was able to ignore the smoke, but now, coming into Escalante feels like driving into a San Francisco fog. In this famously scenic place, with buttes and mesas in the distance, I can't see more than fifty feet ahead. A blurry world. Ghostly, with the oranges and yellows of striated rock visible but not the rock walls themselves.

My time in Eden was nice but the smoke reminds me that we are a species under threat. One of many species of course, all courtesy of us. No need to debate with any locals whether human beings created this or not. They can see (or not see) as well as I can. This is our reality. The next day as I leave Grand Staircase and stop at the top of Boulder Mountain a sign tells me I am looking down at the Aquarius Plateau, the Henry Mountains, the Little Rocky Mountains, Circle Cliffs, and, far off, Navajo Mountain. But all I see is smoke.

———

As I drive west a new phrase, "heat dome," is on everyone's lips. The phrase basically describes what happens when warm oceanic air becomes trapped in Earth's atmosphere by high pressure, but perhaps it is easier to picture placing a metal dome over a piece of meat on a grill. We, and the land and other animals, are the meat. Whatever you want to call it or however you describe it, people keep saying it has never been so hot, and that subjective impression is backed up by the facts. The National

Oceanic and Atmospheric Administration reports that "over a six-day period during the middle of June 2021, a dome of hot air languished over the western United States, causing temperatures to skyrocket. From June 15–20, all-time maximum temperature records fell at locations in seven different states (CA, AZ, NM, UT, CO, WY, MT). In Phoenix, Arizona, the high temperature was over 115 degrees for a record-setting six consecutive days, topping out at 118 degrees on June 17."

Extreme heat in Phoenix is to be expected, but records have also been broken in Salt Lake City, where it hit 107 degrees, and in Billings, Montana, where high temperatures averaged 100 degrees for six straight days, topping at a record 108 degrees. In Torrey, Utah, which rarely hits 90 in the summer, it is over 100 degrees and the juniper trees have begun to die. There is no relief by the ocean either. A much-shared article describes how tens of thousands of clams, mussels, sea stars, and snails were found *boiled to death* (holy shit) on a beach in Vancouver during that country's record-breaking heat wave. In the Pacific Northwest, Portland breaks all records, while Seattle, which from 1894 to the present had only recorded *three days* when the temperature reached 100, now reports three days *in a row* of hitting that number. Glaciers are melting atop Mount Rainier, where summer temperatures are normally near freezing. In July the thermostat atop that 14,409 foot peak reaches 73 degrees Fahrenheit.

Then the fires. Technically, there has at this point been one fire season on record with more actual fires recorded, a decade ago, in 2011, but it looks like that record will soon be surpassed. (It will.) The early heat and late monsoon season have left people wary. (Monsoon season is what people in the region call the period of summer rains where almost all of the year's meager rain falls.) Smoke is everywhere, much of it blown across the interior from the coast. By late July the National Interagency Fire Center's situation report lists a total of 37,009 wildfires across the country that have burned almost 3.4 million acres. The region remains a blur as I drive. My driving is part of the problem; I am a contributor to all this. Does bearing witness justify my journey? No.

Salt Lake City was originally the westernmost destination of my trip, but once I get there I decide to keep pushing it, following the smoke. On the morning of July 16, I cross into a state that looks, perhaps more than any other, like the landscape many scientists imagine in our arid future. If I'm really trying to picture 2063, I could do worse than Nevada.

As anyone who has ever traveled west from northern Utah into Nevada along Route 80 knows, you must first pass through a world of salt. Not just the lake that gives Utah's capital city its name, but the clumped white mountains along the road. This roadside ground salt is thick like ice. I pass the Morton Salt factory, imprinted with the giant logo of the girl with her tilting umbrella, and then, as I near the border between the two states, the Bonneville Salt Flats themselves, home of so many land speed records, its salt flats stretching out forever like a field of drifting snow. Salt is one thing we should still have a surplus of in the future, and here, where the ocean used to be, it is everywhere.

Nevada is next. I have always found this state underrated when it comes to beauty, but it is less beautiful today with the smoke rendering the mountains mere shadows of themselves. Shimmering blue glassy mirages rise on the road ahead, as I point the car toward the great smoky wall of California. After hours and hours of driving through the brown of Nevada, passing too many dust devils, too many signs that read "Do not pick up hitchhikers, correctional facility ahead," and more than a few billboards advertising the coming chukar tournament in the town of Battle Mountain, I finally begin to curl upward along the Truckee River toward Reno. Two sunsets glow in front of me. One actual one and the other, further south, something that looks like a sunset but isn't. The reds and purples and swirling blacks, like a great bruise, of the southern pseudo-sunset are the reflection of a fire, not the fire of the sun. Lit up below, the whole landscape is crisp and brown and friable, despite the green of the cottonwoods along the river.

It is getting dark by the time I climb the Sierra and come over Donner Pass and down into California, and so I pull over in some national forest land, lay out my pad, and try to sleep. It is quiet when I arrive but then headlights come shafting through the trees. I have a neighbor. A rapist? A murderer? A friend and fellow nature lover? I consider sleeping in the back of the rental van but it is too hot so instead I dig out a bottle of Xanax, a prescription we got to give our dogs during thunderstorms, and eat a quarter tab. If my neighbor decides to kill me he won't have much of a fight. Thanks to the doggy Xanax I sleep like a (drugged) baby, blanketless on the duff below the Douglas firs. I wake at first light below the towering firs: a morning of such stunning beauty here in the California forest that it almost makes up for the lack of coffee. I wave to my sweet-faced, non-murderous neighbor and his golden retriever on my way out of the forest.

Westward I go through the burnt golden fields of California until the land starts to rise again, like a ridge atop a stegosaurus's back. The ridge leads to Paradise.

Paradise, California, the scene of the devastating Camp Fire of 2018, is again under threat of evacuation, this time from the nearby Dixie Fire (which will prove to be the largest in California's history). I can taste smoke on my tongue as I drive along the Skyway, the same road where families fled three years ago while flames curled up around them. On a parallel ridge, to the east of town, runs a ghost forest of blackened trees. The town looks abandoned this morning. Empty lot after empty lot. A hand-painted sign advertises "Stump Grinding," and it occurs to me this must be a popular business here. Three years after the fire there are few healthy trees but still plenty of stumps and brush piles on every lot.

Nothing looks open and I am hungry. I pull into a Food Mart at the Phillips 76 gas station.

When I pay for my iced tea I don't plan on asking the tall, gaunt woman behind the counter directly about the fire, but it comes up when I wonder out loud if there is anywhere to get breakfast.

"There used to be lots of stores and lots of places to have breakfast before the fire," she says. "Now there's just one."

I ask her if she lived here during the fire and she tells me matter-of-factly that she did and still does and that she watched her house burn down. They tried to save it with hoses but then a big powerline fell on the roof and that was that. She now lives in a motor home on the plot next to the burnt outline of her house.

She gives me the directions to Debbie's Restaurant.

I pull into Debbie's and walk under the blue awning and take a seat at the counter. A cheerful blonde waitress takes my order of ham and eggs. There is only one other person at the counter, three seats down, a man in work overalls who looks like a bespectacled version of Dusty Hill, the bassist for ZZ Top. We both keep silent at first, just sipping our coffee, waiting for our food, though it seems clear we will soon enough start chatting. Once again I don't want to lead with the obvious—"How about that big fire you folks had?"—but on the other hand, with the smoke thick and nearby homeowners evacuating, is there really any other topic?

So I mention the smoke.

"Oh this isn't bad," he says. "When it's dark in the daytime, that's when you know it's bad."

I ask about the current level of nervousness in town.

"It all depends on the wind," he says in reply. "Right now it's blowing northeast."

"That's good, right?"

"Good for us. Not so good for the folks up in Almanor."

He takes a sip of coffee and continues without prompting.

"Last year we had a fire break out twenty-five miles east of Oroville. And they were watching it, saying it is under control. Then the east wind came in again and damn near burnt out Oroville. A forty-mile-per-hour wind. When it's still like today, sure, you can get it under control. But when that wind kicks up it suddenly moves. They say that during the Camp Fire it was traveling four football fields a minute."

Most of what I knew about Paradise before today came from news clips and YouTube videos, one of a man and his daughter escaping from town while driving a road where fire kicked and spat up from both sides of the highway. Now, before our food comes, I get an education from my bearded neighbor, who introduces himself as Doug Hays.

The fire started with a falling electrical line from a transmission tower. That was at 6:15 in the morning. One of the most surprising things about the fire is that it descended the steep gorge to the east, jumped the West Branch Feather River, and then climbed toward town. That hadn't happened in a hundred years. What had started the burning in the town itself wasn't the fire proper but a rain of superheated embers, which feasted on pine needles, duff, leaves, and the houses themselves. During the half-year before the fire they hadn't even gotten an inch of rain.

The road they had all driven out on was the Skyway, which ran right outside of Debbie's along the ridge. Doug lost his house and wasn't allowed to get back to see his property for three months, until January.

Doug has been homeless since the fire, but he has also had a roof over his head. When he asks me where I am from I mention that I am originally from Massachusetts, fearing the standard Mass-hole reaction. But he says he has been staying with a family from Massachusetts who live in nearby Chico while he rebuilds his house on the same property where it burned. The family has been incredibly generous. But, he adds, there are plenty of people up in Chico who are less so.

"Survivor's guilt is a real thing," he says.

I nod. People all deal with disaster in their own ways.

Doug mentions that his new house, almost completed now, is threatened by the Dixie Fire.

Another diner takes a seat a couple down from me and soon joins our conversation. He could be Doug's fellow ZZ Top band member, a look that has become increasingly popular during the pandemic. His name is Phil and he lives closer to the latest fire, the Dixie Fire. He evacuated just two days ago.

I no longer feel bad about talking directly about fire. I tell them the story of Ken Sleight's Quonset hut, how he had been a kind of archivist, what he lost.

"That's what I was in my family, too," Doug says. "I still do it, still collect everything, even though I know it means nothing and can go away. You can't control it in the end; it means nothing."

I am not sure what "it" is but I don't ask.

"I lost everything," he says. "Even my guns."

He explains. He kept his rifles in a safe, and when the fire heated up they shot off and put holes in the metal. Then, after the fire, the rains came and rendered his guns useless.

I mention the waterproof safe that I bought for hurricanes. Doug doubts it would really work.

"Nothing is safe," he says, and then laughs at the pun and revises his sentence:

"Nothing is anything-proof in the end."

—

A couple months before fire tore through Paradise in 2018, Hurricane Florence hit my adopted hometown of Wilmington, North Carolina. I had been studying climate change and sea level rise for almost two decades, but I was about to be taught a less theoretical lesson. We were not new to hurricanes, having experienced a dozen or more since moving south. But this one, we were told, was different.

It's coming, they said. *It's coming. No, no, it's really coming this time.*

We were not so sure. For many years, living on our spur of the southern North Carolina coast, we had been fooled by the Cantore who cried wolf. We fell for it every time. The Weather Channel's Jim Cantore stood out there, the brave weather forecaster, on the sand below our piers, and warned us of peril. He and his ilk foretold our doom, though their stern warnings were slightly subverted by the fact that you would always see, at least on the beach near where I lived, surfers trotting out to the water behind him, waving to the camera. Once—and I swear this is true—just as our local Jim Cantore-wannabe was telling a TV camera that *No one should go out in the storm* (except for him of course) I saw a woman behind him pushing a stroller, thinking perhaps that her newborn might like the feel of the wind on her face. As you can imagine, this somewhat undermined the point the reporter was so emphatically trying to make.

What I'm trying to say is that in Wilmington, North Carolina, we were a different breed of skeptic. We were the bull's-eye that was never hit. The land of close calls but always misses. So we regarded the warnings with a grain or two of salt. We would board up, we would pack, we would evacuate, we would panic, and then...*pffttt.*

They promised, those priests of the Weather Channel, they promised us that this time there would be no *pfftting.*

It was Tuesday, September 11, 2018, two days before the first bands of Florence hit. I was standing under Johnnie Mercers Fishing Pier, where, after a nice swim, I spent some time observing the Cantore crowd and their hangers-on. It was a meta scene: people using their phones to film the cameramen who in turn were filming the handsome weathercasters who were pointing out at the calm sea, uttering dire warnings (apparently untroubled by false prophecies from their pasts). This was the current world of climate change in microcosm. The seas will rise, the media prophets warn, while we point and click with our cameras. Does anyone here believe—*really believe*—that the island we are standing on will be underwater by 2063?

And yet even we, deep skeptics that we were, were worried. On my way back from the pier I drove along the beach, past people nailing up plywood, before arriving at the apartment where we had once lived and had ridden out a storm or two, the apartment near the ocean where my daughter spent the first six years of her life. My former neighbors, Robert Boyce and Star Sosa, were cleaning out their ground-level garage for fear of storm surges, something they have never done before.

"This one feels different," said Star, who was a twenty-year veteran of Carolina storms.

Robert, an almost daily long-distance swimmer out in the Wrightsville Beach waters, was worried, too.

"I've never felt the water warmer. Never."

We were skeptics but we were not stupid.

Nina and Hadley had already evacuated with one of our yellow labs and the cat, and early the next morning I would take our second yellow lab, Missy, and follow them.

In the meantime I, along with thousands of my neighbors, went about the work of evacuation. This meant bringing everything inside, boxing books and papers in plastic containers, carting more books and papers and Hadley's baby photos to my office at school (though after the storm I would learn that the roof of our school building was only rated for ninety mph winds). It is a strange business, leaving your life behind. There is only so much you can carry in a car or pack in places you pray are safe. For instance, I had about sixty journals I had kept since I was sixteen that I would hate to lose. I jammed them and everything else I could into closets and nailed them shut, knowing this wouldn't do much if a tree blew a hole in our walls.

We were leaving our homes to their own devices. We assumed we were coming back. But we didn't *know* we were. More and more, people across this country and the world are leaving and when they come back their homes, due to wind, water, or fire, are not there.

"Firm ground is not available ground," wrote the poet A. R. Ammons.

Amen.

It was mid-morning the next day when I evacuated. I expected massive traffic jams but most of my neighbors had already left, and Missy and I flew up I-40 with only a few delays. The journalist in me felt guilty. Wasn't it a dereliction of duty to be driving *away* from the storm not *toward* it? There was virtually no traffic going in the other direction, east toward Wilmington, but I imagined the few cars I saw belonged to my fellow writers, journalists, filmmakers, and photographers. *That should be me*, I thought. Instead I was running away.

We would be staying at my sister's house, though the path of the storm was still uncertain, and we had talked about retreating farther inland. While we were there Hadley and Nina would be staying on the top floor in my niece's room, and I would take over my sister's husband's study. He was a chaplain at a southern university, a university with Baptist roots, but he is a secret Buddhist. I dropped my bags and books between meditation pillows and statues of the Buddha.

Once we were unpacked I called my neighbor Tony back in Wilmington. He told me that my writing shack down by the creek, where I wrote in the evenings, was already flooded and lifting off its foundation. After that call, it was hard to get a read on what was really happening with the storm, despite all the news coverage. Tony's phone went dead, and there was a sense of Wilmington being cut off from the world. The Port City, as it is sometimes called, was delicately stepping back in time into an earlier century, one without electricity or passable roads. It was a time without enough water, too, though plenty of water in other ways. The rain fell relentlessly.

When the power came back on, and his phone was working again, Tony sent me a picture of the shack with six feet of water in it. More trees had come crashing down in our yard, though so far none had hit our house.

Others were not so lucky. A former student of mine, David Howell, evacuated with his wife and two children to Charlotte, four hours inland, on the Tuesday before the storm. His wife, Jo, sent out a text to friends on Friday, a day after the storm hit, saying it looked like all was okay with their home back in Wilmington. As David tells it, it was less than a half hour later when his neighbor called.

The neighbor hemmed and hawed and apologized to Jo until she asked him to get to the point. The point, it turned out, was that the huge water oak in their front yard had fallen and crashed into the middle of their house, creating a hole in the roof above their daughter's room. There was no way for the neighbor, or anyone else, to do anything about the tree, so for the next two days, the rain fell into the house and the winds blew through it and a possum took up residence, no doubt seeking shelter from the storm. Since they lived on the creek and had been worried about flooding, they had moved everything upstairs, which was now open to the weather. A microburst also picked up a smaller tree and tore more holes in the house.

They had left their house to its own devices and it had failed them. The five days before they could get back were days of sheer panic. *What's going to happen to our stuff? What just happened to our lives?* And also anxiety about getting in front of the computer and filling in an insurance claim.

"When you lose a house, you lose a family member," David told me later. "It's like a mother. The ultimate protector."

Though they would have no place to stay when they came back, they felt an increased sense of urgency to return—while knowing there wasn't much they could do once they got there. Finally, they made their way back, zigging and zagging through

North Carolina, avoiding the many flooded roads. They dropped their kids off with a friend and went over to walk through the house. The possum had shat and pissed everywhere, and the walls were wet and dripping.

"We just lost it," David said.

After an hour or two of despair, they called their friends and coworkers and cleared out of a house they had lived in for six years in just two and a half days. They'd lost most of their stuff, including everything their daughter owned. All of their belongings were piled up on the street along with the trash and the tree limbs.

"That's another thing," David says. "I loved that tree, that big water oak, that fell. And there is a sister tree of that tree in my backyard. I don't want to do it, but I'm going to cut that tree down."

As bad as it was to lose a house, even they were relatively lucky. At least they had their lives.

The story on everyone's mind was that of a woman who, holding her baby tight, was crushed when a tree fell through her roof. They both died immediately but the husband survived. He was bawling when the ambulance, somehow making its way through a forest of downed trees, arrived. Rushed to the hospital in critical condition, he had truly lost everything—wife, daughter, home—thanks to one burst of wind, one falling tree.

For the next two weeks the rain wouldn't stop. Some people thought they made it safely through the storm, but then they saw the water rising toward their doorsteps. The storm dumped ten inches of rain, then twenty, but still kept going. Over thirty inches of rain would fall and for the first time on record our annual total would top one hundred inches. The world was drenched through. Permeated. Tony sent us a picture of the lake that was our backyard. Fallen trees floated on that lake along with pieces of the shack. We had given him a key before we left and he took a tour of our home. It looks good so far, he told us.

No leaks. (It wasn't until a month later, thanks to inept roofers, that we would experience interior rain.)

Many of our friends had stayed, and we worried about those we had not heard from. There was some vanity in all this: on social media, and regular media, Wilmington was suddenly a star. Some friends, who had not suffered too great losses, would look back on this time nostalgically. But most admitted to their misery.

After the winds and days of rain the humidity returned, the blazing heat. Even under normal circumstances, Wilmington in September is your town on the hottest day of the year. Food rotted in powerless fridges. People were hungry, unshowered, uninformed. Some had not had power for days and there was no word when it would return. Primitive times. The few stores and restaurants that were open became neighborhood commons. People needed to get out of their stifling, leaking houses. People needed to see other people. People needed to tell their stories. Stories of disaster: the tree that fell right through our living room, that split our house. Stories of near misses: the man who decided at the last minute not to sleep in the room he deemed safest and woke up to find that an oak tree had bisected that bed.

Tales of survival, yes, but also tales of ruin.

—

Florence made landfall in North Carolina on September 14, 2018. Not quite two months later, on November 8, the emergency workers that we had come to think of as "ours" headed west, as did the eyes of the country, to Paradise.

After my breakfast at Debbie's Restaurant, I follow Doug's directions and drive up the Skyway. This is where the fire climbed the ridge and it is hard to be here without thinking of the videos of that night. The woman praying out loud with her family as the flames licked and blew toward her car from the road's edge. The brave father driving through the same flames telling his young daughter in a calm voice: "We're not going to catch fire. We're going to get out of here." The relief anyone watching that video felt when they drove clear and the girl yelled "You did it!" before her father corrected her: "*We* did it."

Eighteen thousand buildings were burned that day and eighty-five people lost their lives. The fire moved so fast there was no way for some to evacuate. It blew up the ridge, uphill being the fastest way for a fire to travel. But even before the fire proper reached town, the winds had blown a thousand sparks, and those sparks had started a thousand spot fires.

I drive north into the next town, Magalia, and pull over on a dirt road. To the east of the ridge that Magalia and Paradise sit atop, the land dives steeply down to the west branch of the Feather River before climbing again to the next ridge. I hike down through a corridor of still-blackened trees. A large rabbit with reddish ears goes bounding off along the path and a deer stands up on the burnt ridge above me. It is kind of beautiful, really: a world of still-burnt bark. Will it ever not smell charred? I come upon two burnt-out cars and a landscape of shattered glass and trash, and follow the burn scar down through the black spindly trees. Blueish smoke rises from the valley. The smell of the new fire mixes with the char of the old. Further down the hazy canyon I watch two acorn woodpeckers, working together, drilling into the pale fleshy wound near the top of a pine tree.

Back in the car I continue north into Magalia. A couple, maybe in their early seventies, are out walking and I roll down my window. I ask where the lake is and the man points down the road and says it is very low. Soon we are chatting and before long, just like with the men in the restaurant, we turn to the obvious topic.

"We didn't even have a chance to fight it," the man says. "We were running for our lives. We stayed until one-thirty in the morning. We could see it coming. We went down through Paradise on Clark Road. Until we got there I hadn't realized the whole town had burned up. I could barely get under the downed power lines in my pickup. There were burnt cars in the road and along the side of the road. Fires still burning everywhere. Our daughter and son-in-law lived in Paradise and we had no way to communicate with them, to know if they were okay. My son lives over on the other side, in the Pines, and we couldn't communicate with him for five days. My mother, who was ninety, her home burned. A neighbor helped her get out but she lost everything."

They have lived here, or close to here, their whole lives. They say it isn't like it used to be. There is no rest from the fires.

The people on TV always talk about rebuilding. The flames have barely been put out, the water barely receded, and there they go, desperately jabbering.

And of course I understand. Every place that faces disaster has to bolster itself with pep talks of resilience. Paradise Strong. Wilmington Strong.

"It's easier when you are younger," the woman says, pointing to her husband. "He says 'If I was only forty years old it would be no big deal. I could do this and do that and get everything back to where it was. But when you're this old...'"

Her husband nods. There is no getting that back.

"I retired and I had a passion for fishing," he says. "I had riverboats and a nice setup on the Salmon River and I was pretty happy. Now that the fire came and burned all that up, I don't even have the energy to start again. To buy new boats, rerig everything."

Now losing a boat might not count for much compared to what will happen, and what has already happened, worldwide. But I understand.

"It's hard to recommit with passion," I suggest.

"Exactly. I don't have it anymore. That is what I did. That was my passion."

We have been talking through my car window but I finally get out and introduce myself. The man's name is Richard Terrano and his wife's name is Joanie. I reach in the car and hand them a copy of a book I recently published.

Joanie thanks me. "I'm a prolific reader. I was worried about my books during the fire." Then she says, "Each disaster is that person's experience." That, I think, is what we miss out on as the camera crews rush from the last disaster to the latest, or when we spew statistics.

People were kept away from their homes for weeks, some for months. What was it like coming back, finding out what was still there and what had been lost?

—

Two weeks after Florence I made my way back to the soaked city, past warped signs and high creeks. Awnings had blown off gas stations and thousands of trees were down. I did not take the main highway that I had evacuated on weeks before. That was because the eastern parts of that highway were now a full and deep river. In the pictures people were posting you would really think it was one, or at least a canal. You could float boats on it but there were no cars driving on it for weeks. This new river effectively cut our town off from the world, making it an island.

When I finally returned to the swamp that was Wilmington, the writing shack, miraculously, was still standing. It seemed to have been lifted and then clumped back down where it had been. Its door and front wall were torn off and an eastern red cedar pressed down on its roof, and mud covered the floor. It was on its last legs, down but not quite out. I could sit inside it still, though it leaned like the lair of the TV Batman's villains. The roof remained snug: the shack lost fewer shingles than our house.

I spent the first morning hammering the fence around our yard back together so the dogs couldn't get out. Some workers I had hired helped me push the red cedar off the top of the shack. The whole root ball was out but we replanted it. Trees were down all over the yard and shingles had been ripped off the roof of the house, but it looked like we truly were one of the lucky ones.

The curfew was still in place. The usual tension in the town was exacerbated by the heat, the scarcity of food. There was looting at a dollar store downtown. The police were called in but it was unclear whether the store's owner had simply opened his doors and let those in the neighborhood help themselves to the food before it rotted. Either way reports of the looting bred fear. My friend Dave went looking for his cat in the neighbor's yard and the neighbor pointed a rifle at him.

During an event like this it sometimes seems as if the poor and the rich could be two different species, at least in their ability to respond to calamity. *Homo benficioius* and *Homo strappus*. The wealthy, at least the smart wealthy, used their money to protect what was most precious: their lives. (I should add that there were plenty of dumb wealthy who chose instead to stay—with their children!—imagining they could somehow protect their property.) But many people had no choice but to ride out the storm.

In the wake of Florence, those who built the most expensive, but vulnerable, homes would receive the majority of the federal aid. They were helped in this by a federal flood insurance program that encouraged people to build on the shore in the first place. It was just what I had seen while reporting on New Jersey after Sandy. Welfare for the wealthy. The ocean had swept the table clean and now the table was being immediately reset. There was the usual reflexive talk of rebuilding. And so, like ants barely pausing after their hill has been kicked over, the residents set out to rebuild exactly where their homes had been built before, that is, right in the line of fire of the next storm.

But there were ironies too. The poor feel the brunt of climate change in ways the wealthy are protected from, but along the Atlantic coast, as in many other places in the world, the houses of those who pay inordinate amounts of money to live by the water are on the front line. It has always been a dangerous gamble to live near the water. In Wilmington there was another twist. The city's most prestigious and tree-filled neighborhood saw the worst downing of trees, while some poor downtown neighborhoods were protected by their relative treelessness. And if Florence brought attention to the economic gap in our city, this sort of primal event also revealed how clearly we are all part of the same ecosystem. That included not just poor and rich, but selectmen and sandpipers, developers and dolphins. All impacted. All intertwined. All part of a complex and messy whole.

My second morning back I was up at dawn, kayaking behind my house down Hewletts Creek out to Masonboro Island to survey the damage. *Homo sapiens* were not the only animals to lose their homes. I quickly saw that the live oak that held the osprey nest near the creek's mouth was down. Great gnarled arms of oak had been upturned, the whole massive root system revealed. Taller than me, it looked more like the root ball of a redwood than a twenty-foot oak. Nine days had passed since the storm hit as I paddled out toward the mouth of Hewletts Creek. I saw several more toppled trees that held osprey nests, but the skies were full of life, and within twenty minutes I had seen not just ospreys but egrets, ibises, great blue and green herons, terns, and a kingfisher who dove ten yards in front of my bow. What I didn't see were other people. There was an emptiness that reminded me of a greater, more deadly disaster. Just as the skies were free of planes after 9/11, the waters were free of boats.

My initial assessment, looking at the homes along the creek, was that it could have been a lot worse. Rain, yes, we had had a preposterous amount, over thirty inches in some spots, and the Cape Fear River hadn't even crested yet. But the winds were not the Category 4 or 5 we had feared. As I looked along the coast, about one in a hundred trees seemed to have been downed. It could have been ninety-nine out of a hundred. I had been in Monkey River Town in Belize after Hurricane Iris hit in 2001 when *all* the trees were down. That storm was a Category 4 and made landfall with 145-mile-per-hour winds, devastating not just the trees but the howler monkey population that made their home in them. This sort of arboreal apocalypse could have happened here. We can't imagine it, perhaps, but it is entirely possible.

Luck is an underrated element in hurricanes, and the lottery of wind is never equitable. *Dawson's Creek* was filmed on the creek behind my house, and fans of the show will be relieved to know that Dawson's dock still stood. Not far from the downed osprey nest, I passed Dawson's house (the actor who played him had bought the fictional house) and the long dock, where a boat was tied up and a fake plastic owl had gone back to work scaring off gulls. I watched a green heron staring down into the water, fishing from the dock.

I paddled across the Intracoastal Waterway to the dredge-spoil island I call Osprey Island, but the nest I named it after was down too. What I found instead was a marooned sailboat thrown up on shore, its twenty-foot-long mast tilted at a forty-five-degree angle. I couldn't quite make out the name of the boat, but the brand was Buccaneer.

What struck me again almost immediately, despite the downed trees, was how much worse it could have been. In the end this was a Category 1. There was no reason a Category 5 couldn't sweep in the next week. For all that our town had dominated the national news, for all the deaths and destruction, if Iris-like or Katrina-like winds had come through here, which they eventually will, we would be telling a different story, a story far darker and more tragic. And so would the animals. Their lives are more obviously closer to the precariousness that defines all our lives. Nests wiped out, marshes batted down, tree homes gone. We were united with them in our loss.

I landed on Masonboro with its eight-mile stretch of beaches, dunes, and marshes jutting out into the Atlantic. The beaches acted as the front lines for any storm approaching Wilmington, but if anything, the island seemed to have done better than the mainland. My usual camping spot looked pretty undamaged, in part because it was so low, nestled in a cup between dunes. "Stay low" is good general advice during a serious storm, and that's something Masonboro Island does naturally. But what wasn't low, at least not after the storm, were the dunes. They were the true front lines, and they had taken a serious hit. Rolling and gradual, smooth and easy during the quieter times of summer, they were now sharp and steep, sheer and clifflike, some twenty feet high. They looked like they had been sliced in half—but they had done their job, protecting what was behind them.

While the island never looks hardy, its dunes blocking the wind and its marshes on its backside allow it to receive and interact with the storms in ways a developed island cannot: sand spilling over the island's interior, the marsh growing behind, the entire island gradually but constantly migrating landward. Supported by its backside marsh, Masonboro still handles storms in the old-fashioned way. Because it has been left alone and undeveloped, most experts point to it as the healthiest example left in North Carolina of the way a barrier island survives and changes during a storm.

The relative success of Masonboro, however, speaks to the dangers of coastal development. When I hiked to the north end of the island, I could see its neighbor, the highly developed Wrightsville Beach, where my family and I once lived and where the Weather Channel's Jim Cantore had camped out during Florence. Wrightsville Beach, too, got lucky during the storm. Despite massive flooding and damage, the relative weakness of the winds allowed most of its houses to survive. Florence, as bad as it was, was not the Big One.

If Masonboro handles storms through a kind of elemental judo—storms pushing sand landward and the island jumping over itself and growing on its backside—on developed Wrightsville Beach this ancient judo technique no longer works, thanks to the three thousand other people who crowd its six miles of shore.

One day during my first year in Carolina I had paddled out to Masonboro and looked back at Wrightsville and our apartment. The island appeared decidedly fragile, and a strange metaphor came to mind. With its flat treeless land and tall buildings, the island looked like nothing so much as a dinner table full of empty plates and bottles after a party, waiting, I thought, for an angry drunk to come along and sweep it clean with his arm.

Florence was not the angry drunk, but of one thing there is little doubt:

Sooner or later the angry drunk is coming.

—

My road back east from Paradise is steep and winding, through the Plumas National Forest along Route 70 and along the Feather River. I pass a couple dozen of the green trucks of the Plumas Hotshots, who are fighting the Dixie Fire. I remember that Orrin Pilkey, the prophet of water, fought fires as a stump jumper when he was a young man.

I can see fire burning on the ridge, the flames kicking up and leaving behind burnt trees like sticks, the red of the flames reflected in the clouds above it, showers of sparks. The fire hungrily devours all that unused fuel, like humans a little in that way.

After another hour of winding through the canyon I pass the site of another major fire, the Beckwourth Complex Fire, where a sign saying "Ground Support" leads down a road to where nine helicopters rest at the foot of a denuded range of purple-black mountains.

—

The tide had shifted for my paddle back from Masonboro. The water was a dark iodine color with gray burbling bubbles. The surface was glassy and, looking down, I could see the reflection of ibises flying above me. But if it was placid and beautiful, it was also scummy. It looked foul, smelled foul—was foul. A great stew of pig excrement, corporate chemicals, debris from shattered homes, human waste, and coal ash had been carried down the rivers to the sea. That was the first time, in any season, that I didn't take at least a quick dip in the ocean during a visit to the island.

A dusky osprey flew overhead, and I knew it had no choice but to take in these poisons. It was the same old story for ospreys. They were the bellwether for the bad old days of the chemical DDT. The story is a familiar one by now: they ate fish contaminated with DDT, and the chemical accumulated in them, at the top of the food chain, through biomagnification. Today it was a different story, but not so different. The fish were swimming through sewage and the ospreys were eating the fish.

Worse was the idea of dolphins swimming through the same water. *Living* in it. It is not just our own nest we have soiled.

I was careful not to let too much of the foul water splash on me as I paddled. While I passed no other boats on the way home, I did come upon two teenage girls, about my daughter's age, on paddleboards. I didn't want to say anything, didn't want to be a *Dad*, and for a few minutes I kept quiet, just paddling in place. But in the end, I couldn't help myself. I imagined them falling off their boards, or jumping in just for fun.

"Girls," I called out to them.

"Sir?" one of them asked. Yes, we were in the South.

"Please don't go in the water. There's a lot of bad stuff in it. Please try to stay out."

"Yes, sir."

I felt bad for scaring them, but I hoped they listened to Sir. The pool of chemicals, waste, and coal ash we were paddling through was perfect for infection. One swimmer would die from it in the coming weeks. The ban on swimming on Wrightsville Beach would last until October. The storm had darkened more than just the water. The beach and sea were no longer a refuge, a place to get away from it all. They were a place where our homes, and our lives, were threatened.

They were a place of danger.

COCKTAILS WITH VULTURES

"The scent of decay is not subtle."

—Noah Strycker

Have you noticed the vultures?

They are everywhere these days. And I'm not being metaphoric.

I first began to take notice, real, deep notice, during these summer stays in Boulder.

Around six on most nights I would head out onto the front deck of the place where we were housesitting, a glass of cold India pale ale in hand, and watch the turkey vultures come home from work. After a long day of looking for dead things up in the mountains they would spiral lazily back to the mangy Douglas fir in front of the house. My workday also done, I was soon sipping a second beer and watching as the birds slowly circled down and landed on the half-dead tree where they would roost for the night. They never settled right away but jockeyed for position, bullying each other off the prime spots, establishing a vulture pecking order, though I hadn't yet read or seen enough to know if this was real or imagined.

The deck was an ideal bird blind, though not a blind at all, since I was barely 150 feet away and the vultures could see me as well as I could them. In a way it was neighborly, though our human neighbor across the street might have felt that the situation was less cozy, with a dozen or more large black birds hovering over her roof like harbingers of death. For me it was perfect, however, and I would scribble down drawings of the vultures on my journal pages, half watching the birds, and occasionally breaching from my own thoughts and preoccupations to toast my new acquaintances.

—

A student recently asked me to recommend a book about climate change.

I suggested he read one that was written in 1941, during an earlier end-time. T. S. Eliot might not have known global temperatures could rise by 2.4 degrees Fahrenheit by 2063, but he sure as hell knew *something* was up. At least that's how I choose to read him.

This, for instance, is from "Little Gidding," a section of Eliot's *Four Quartets*:

> *Ash on an old man's sleeve*
> *Is all the ash the burnt roses leave.*
> *Dust in the air suspended*
> *Marks the place where a story ended.*
> *Dust inbreathed was a house-*
> *The walls, the wainscot and the mouse,*
> *The death of hope and despair,*
> *This is the death of air.*

The place where a story ended. The death of air. These were not sentiments you are likely to hear on CNN, or espoused in the Senate, but I stand by them.

And by this:

> *The only wisdom we can hope to acquire*
> *Is the wisdom of humility; humility is endless.*

The houses have all gone under the sea.

—

The vultures settle into their roosting tree by seven at night and are, by bird standards at least, relatively late risers the next morning, which means that they just sleep and sit around for a good twelve to thirteen hours every day. Then in the morning they at last, in the immortal words of Jackson Browne, "get up and do it again," riding the thermals to the mountains. Spiraling upward.

One morning a couple summers ago I walked down from our temporary house, past the cemetery, to the college center called the Hill to get coffee for my wife and a vegan donut for my daughter. A dozen vultures hovering over the cemetery seemed a kind of morbid joke at first, but I stopped to watch anyway. I could see them circling, trying to get heat under their wings, and I wondered if the cool morning made it harder. As usual, they were rising late—it was already nine—but gradually they began their daily ascent, moving up in the world, rising and rising, describing larger circles that took them closer to their destination. That destination was the top of the nearby mountains, but also the road that ran up those mountains, and I wondered if they might sometimes stop in mid-commute to snack if they came upon some roadkill. My own daily ascent would come later when I biked up into the same hills. Earlier that week I had been biking back down when a fox with a mottled black and orange coat ran across the road with a squirrel in its mouth. There was lots of death on those roads and in those hills, which meant that for the vultures it was a land of plenty.

—

I have lately made it a practice to read all the climate books I can get my hands on. Which isn't exactly a blast. Two recent ones I read were *The Uninhabitable Earth* and *The New Climate War.* The second, written by Michael E. Mann, acted as a corrective to the first, written by David Wallace-Wells. Mann criticizes Wallace-Wells for going full doomsayer, cherry-picking the worst of the worst reports and therefore sensationalizing a crisis that on its own is sensational enough.

I understand Mann's concerns and have often argued the same myself. I am made uneasy by an argument that feels like a panic attack. But maybe I'm coming around. I seem to be warming to the notion of a possible apocalypse.

During the weeks after the attack on the United States Capitol in January, I thought a lot about the power of words. This is not unusual; I'm a writer after all. You may think I am referring to the power of words to distort, since that was what was making the headlines back in the winter of 2021. That was part of it, but I was also thinking about the way that words can inspire and bring new things into the world. And the way they can fail to inspire.

In *Four Quartets*, T. S. Eliot wrestles with how we face our ends, personal and beyond, but he also considers how we use words, and what language can and can't do. He helps us imagine both a new language, and the difficulty, if not the impossibility, of creating a new language:

> *Trying to learn to use words, and every attempt*
> *Is a wholly new start, and a different kind of failure*
> *Because one has only learnt to get the better of words*
> *For the thing one no longer has to say, or the way in which*
> *One is no longer disposed to say it. And so each venture*
> *Is a new beginning, a raid on the inarticulate,*
> *With shabby equipment always deteriorating*
> *In the general mess of imprecision of feeling,*
> *Undisciplined squads of emotion. And what there is to*
>   *conquer*
> *By strength and submission, has already been discovered*
> *Once or twice, or several times, by men whom one cannot hope*
> *To emulate - but there is no competition -*

*There is only the fight to recover what has been lost
And found and lost again and again: and now, under
  conditions
That seem unpropitious. But perhaps neither gain nor loss.
For us, there is only the trying. The rest is not our business.*

"Raids on the inarticulate" are never easy.
You had better come well-armed.

—

So how do we think about the fate of the world?

Generally, we don't. We avoid it.

Our own lives on the other hand...While the end of humanity may scare us, can it really compare with the end of *me*? With the unimaginable fact that I, of all people, may someday not exist? And finally can we truly fight for a world with us not in it? Can we imagine beyond ourselves?

We are told that only narcissists put themselves above the world, but truly who doesn't care more about self and family than the fate of the planet? Not that we don't try to stretch our minds and extend ourselves outward. But we are animals trained by millions of years of evolution to fight for our tribe, for our innermost circles first. How to take these tribal minds and truly fight for the world? I don't mean pay lip service and virtue-pose and claim to care, but really dedicate ourselves to that fight? For anyone who truly does it has to be tied to ego in some way, at least at the start. To self-worth. To self. That is the only road, for most of us, toward anything like selflessness.

My tribe of artists, writers, and poets are particularly susceptible to narcissism. Many of us have pat answers when we are asked why we do what we do, and I'm no different. Usually my reply is a soft one peppered with the word *love* (as in "I love my work" or "labor of love"), and sometimes if I'm feeling literary I throw in the old Robert Frost quote they drag out at graduations about "uniting my avocation with my vocation." When less inspired I resort to some well-worn variant of "I write because I *have to*." Whatever. These are nice answers, all of them, but the truth is it isn't something I have thought about too deeply, and if I did I suspect the true answers would be significantly darker and less verbal. A series of angry but self-asserting grunts maybe. A howl or a pounding on my chest.

In fact, as I turn toward rooting out a truer answer, I can't help but suspect that my current obsession with words is entangled with my first obsession, the one that plagued my late childhood and early puberty. I was a fairly normal kid, if there is such a thing, good at sports and not ugly, and though my brain teemed with the usual childhood insecurities, many adults seemed to be fooled into thinking that I was a confident and relatively healthy-minded young man. They were wrong. Below my ready laugh and not unpleasant exterior lurked a dark, obsessive fear of nonexistence.

I think I was around ten when I had my first crisis of being. This was not a Sartre-like intellectual consideration of nothingness that provoked mild nausea, but a visceral sensation that induced something close to real madness. A sudden and overpowering sense that there was nothing in the world: that the world and, more importantly *I*, didn't exist. This sensation provoked something wild and strange in me, a pure panic reaction that the child psychiatrist who later observed me described as being like "an LSD user on a bad trip." Not the kind of thing young parents want to hear about their ten-year-old.

It all started when my dear dog Macker died. Macker, a collie who looked like Lassie, had been my lifelong companion, brought home from the pound the same month I was born. His death was an ugly one: he ate some of the salt that was used to melt the ice on the roads in Worcester, Massachusetts. My mother found him frozen stiff in a snowbank in our backyard. As anxious as I was about my beloved dog's whereabouts, neither my father nor mother told me anything about his death for almost a week. They waited because they were worried about my reaction, and as it turned out they had good reason to worry. I flipped out when I finally heard. I couldn't believe that this thing—this horrible, cold ending—had happened to my best friend. Even more appalling, as the days passed, was the notion that something similar might one day happen to me. Lots of kids start to worry about death around this age, I understand, but not many begin to truly obsess over it. Lying in bed at night a thought—or more accurately a series of thoughts—would grip hold of my mind. These thoughts would comprise my first full-blown obsession.

I called it "the feeling," and though it began with thought it ended in outright panic. It was a little like walking up a mental staircase where each step was more frightening. It all began with the fact of death, but that was just the first step. Next I'd imagine how it felt not to exist at all, and it felt like nothing, as well as I could imagine nothing. But with some effort I would manage to put myself in this state of not-being, and with that I would begin to sweat and grow nervous. At that point I wasn't quite terrified. Not yet. Terror was the next step and with it I left logic behind. In my solipsistic manner I reasoned that if I didn't exist then nothing else did either. It was about then that "the feeling" usually took over.

It is impossible to exaggerate the sensation of terror that came over me as I ascended the next step. Suddenly, I was certain at that moment that *I did not exist* and, equally if not more terrifying, *the world did not exist.* I was nothing—not even something that might have ever existed. I pounded on my chest to remind myself that I was solid, but it did no good. I reminded myself that I had to exist because I was the one thinking these thoughts, but that didn't help either. I felt like an imaginary wisp, a fleck of nothingness, a passing thought in someone else's mind. Worse, it was as if I was merely a fleeting second in someone's dream, but I was even less than that: the dreamer did not exist.

While my logic might have been faulty, the power of what I felt was undeniable. I ran through the house like a lunatic. I had a room on the attic floor and when "the feeling" hit I would sprint down the stairs, screaming. Once I picked up a painting and almost smashed it over a chair—after all, it was *nothing*—only holding back at the last second.

Usually the first person I encountered was my sister, Heidi. Heidi and I were close, but it still surprised her when her lunatic brother came charging down the hall and hugged her tightly in his arms. I needed to hold something solid, to prove to myself that something did indeed exist, and Heidi usually existed. But even squeezing Heidi like a boa constrictor didn't make her feel real to me. I would sprint off to find my mother, then clamp onto her as tightly as I could, trying to find something *real*. But it didn't work, it never worked. Though she felt substantial enough, I knew the truth, knew she was just another illusion.

I tried desperately to explain how I felt as if that would somehow reduce the horror.

"Nothing is nothing!" I yelled at first, getting my words mixed up in my panic. "Nothing is nothing!"

"Of course nothing is nothing," my mother said with a gentle smile.

Couldn't she see the awful truth? Why was she smiling? Was she *crazy*?

"But don't you understand: nothing is nothing!"

She smiled again and I realized my mistake.

"Everything is nothing!" I screamed.

I wanted these words to strike her with the force of revelation, just as they'd struck me. She, after all, was the person I was closest to, the person I had *come from*. But I couldn't make her see, no matter how I tried. She would pat me on the shoulder and assure me that it would be all right. But it *wouldn't*, I knew that, knew that nothing would ever be all right again.

One time I had the doubly unpleasant experience of having "the feeling" strike when my father was around. While usually intimidated by him, that day, overwhelmed by my obsession, his presence hardly mattered. He was just another whiff of nothing (though, granted, a more substantial whiff). I charged around, throwing things, screaming, "EVERYTHING IS NOTHING!" (I'd gotten the wording right by then.) At first he yelled back, angry, until it finally dawned on him that I was, at that moment, almost completely out of my mind.

He looked at the frothing maniac who was his firstborn.

"Calm down, David," he said. "Just calm down."

He rapped his knuckles on the coffee table to show me how hard it was.

"This isn't nothing," he said. He touched his own burly chest. "I'm not nothing."

He had the right idea, I'll give him that. But it didn't help. I stared for one intense second at his puzzled, apelike face. Then I sprinted off down the hallway, screaming and yelling.

"David!" he yelled. "David Marshall Gessner you come back here this instant!"

I didn't come. He found me cowering out in the garage.

"Look, David, just try and calm down—"

"You don't understand!" I yelled.

He'd had enough.

"I understand one thing, my friend. I understand you don't see how lucky you are. Lucky to live in this house, to have food on the table, to have two parents. Do you realize there are millions of children starving and dying of disease?"

I looked up at him as if he were the crazy one.

"I'd like to be starving or dying right now!" I screamed.

And I believed it! Believed that I alone was cursed with this awful understanding, an understanding that made starvation and disease look like child's play.

I ran away from him, down the street into the neighborhood. I couldn't listen to him and his logic. Didn't he understand? *Everything was nothing!* I did not exist!

"The feeling" didn't happen every day, but it happened often enough to cause my parents serious worry. It wasn't particularly pleasant for them to suddenly be contemplating turning in my Pop Warner uniform for a straightjacket. Though I tried to convince my mother it would do no good, she finally dragged me off to see a psychiatrist. With the first one I got nowhere, but the second was different. I was twelve or thirteen by then and I folded my arms as I sat up on the couch, convinced that mere words couldn't change "the feeling."

"I won't try to 'cure' you," he said. "All I'd like to do is teach you how to relax yourself. So that when you have your 'feeling' you can use these techniques."

He was a bearded Black man who wore a tweed coat and smoked a pipe. He seemed very calm and wise, but deep down I knew mere "techniques" didn't have a chance in hell against "the feeling." Still when he asked me to, I stared at the spot on the wall.

"I'm going to hypnotize you, David, but it's not like in the movies. Anytime you want you can come out."

I nodded and counted backward. The plan was that, while under hypnosis, he would teach me how to hypnotize myself. Then, when the feeling came, I could calm myself down with self-hypnosis. I went along with him, knowing full well that I wasn't about to close my eyes and count backward when the real thing returned.

And, just as I guessed, it did return. But then, slowly, mysteriously, it faded. Maybe I was just getting older, growing out of childhood logic, or simply getting used to the idea that I was "nothing." By the time I got to high school I was only experiencing it about once a year, then not at all. I never believed I was cured (and I still don't, I suppose), but, on the other hand, I have to admit it stopped. Whatever the reason, "the feeling" finally faded away.

—

A good birder can spot a vulture from well over a mile away. Birders use an old military term, GISS, which is an acronym for *General Impression Shape and Size*, to describe the general sense, or gestalt, of a bird when seen from a long distance. GISS works particularly well with turkey vultures. Even from far off you can spot them quickly. Their flight is unstable, relatively wobbly compared to a hawk's, and their wings form a V shape, or dihedral.

I said that vultures are everywhere these days. I'm not trying to be melodramatic or funny. It's true. If you, like me, are always checking the sky, they have become the default bird. Whether on Cape Cod or in Colorado or Carolina, those shaky black Vs are a constant.

They even have a direct link to climate change. John Hay said he never used to see them on Cape Cod in winters, but now they were everywhere. His theory? "In the warming world, winter roadkill and other kill are no longer frozen."

During my summers in Boulder I have learned that a turkey vulture's day is easy enough to observe from the outside. The ones I watch wake, lounge, defecate, and then slowly rise into the hills to soar and search for dead things. They lack the vocal organs of most birds and the only sounds they supposedly emit are the occasional irritated hiss or whine. I have noted a kind of chitting sound too. Other than that, they are silent.

They are death eaters, though not of the Harry Potter sort. In his essay "The Buzzard's Nose," Noah Strycker translates the turkey vulture's Latin name, *Cathartes aura*, as "purifying breeze" and says, "It is an apt scientific name, as turkey vultures serve the unenviable task of cleaning up the world." Vultures evolved to be able to digest almost anything, including rotting flesh and animals that have died from disease, and then to excrete shit that is completely sterile. Really.

If vultures could contemplate climate change, and who is to say they can't, they would no doubt be less anxious about it than human beings.

"Vultures have evolved to survive on things that would kill us," Strycker writes.

—

Vultures, really? Why not just write about a dark-cloaked guy with a scythe?

I apologize for my morbid preoccupation. Like a lot of us, I have had a rough couple of years. They have been years of loss, and the climate disasters of this summer seem to echo the personal losses. The old guy in the Eliot poem is starting to feel a lot like me:

> Old men ought to be explorers
> Here or there does not matter
> We must be still and still moving
> Into another intensity
> For a further union, a deeper communion
> Through the dark cold and the empty desolation,
> The wave cry, the wind cry, the vast waters
> Of the petrel and the porpoise. In my end is my beginning.

"The feeling" is gone, a childhood delusion, but sometimes I wonder if it is just buried. Lately, I have been considering the art of not thinking about things. Maybe that is the main skill of adulthood. Maybe that is why those young activists, nagging us about the end of the world, bug us so. The temperature is 120 in Phoenix? I'll wait and play golf tomorrow then.

Repression can be an effective tool. Who could get through life without it?

It works, yes, but on some level we also know that all we have really done is fashion a pair of custom-fit blinders. Which, again, is not so bad. The best of these blinders come with something called a *purpose*.

In this way death, as I'm sure the vultures would argue, isn't all bad. In his brilliant book *The Denial of Death*, Ernest Becker puts forth the case that much of our energy, much of our creativity, much of our *life*, comes from our attempt to deny the essential fact of our existence: that that existence will end. Whether we consider ourselves life-affirmers who claim the fear of death has no hold on us, or "realists," who admit to living in death's shadow, we are all, according to Becker, both terrified and propelled by our not-so-happy endings. We throw ourselves into frenzied attempts to fill up the nothingness with "something," hurling our objects of work or art—our creations—into the void. At the same time we often try to make a *name for ourselves*, knowing that the worst thing we could be is a "nothing." Speaking not just of artists but of human beings in general, Becker says that in the face of our realization that we are nothing, we fight to stand out, to be *something*, trying to build a narcissistic shield around ourselves that keeps death out. We see this need to win, to be first, most obviously between siblings as children, but it is also obvious enough in adults. Becker writes: "But it [narcissism] is too all-absorbing and relentless to be an aberration, it expresses the heart of the creature: the desire to stand out, to be the one in creation. When you combine natural narcissism with the basic need for self-esteem, you create a creature who has to feel himself an object of primary value: first in the universe, representing in himself all of life."

The next step is finding obsessions that reflect ourselves and our shining narcissism. For artists, obsession obviously comes in handy. It not only gives us the energy and power to create the artistic object, but it fills up our minds in a way few other things could. But can obsession fill the death hole? Of course not, though maybe it is out of nothingness that we all begin to create. If the world doesn't exist then we will make our own world. Maybe all this fever of creation, this need to be special, this frenzy—what Thomas Wolfe called an "enormous task of excavation" of self—comes at least in part out of the terror of pure emptiness, the terror of the end. The need to fill the void, to make something out of our vast sense of nothing. Extreme fear of oblivion creating extreme creation. We hurl ourselves against the death void.

In the end, Becker acknowledges, death wins. "All paths of glory lead but to the grave." The least hardheaded moments in Becker's book are when he reasons that, since everything man does is an illusion, why not pick the best, the highest, illusion. This is Becker's somewhat convoluted path back to the spiritual, to god. But there is another path to take here, another choice. What if we acknowledge that all our dear passions—ambitions for fame or love or spirituality—are illusions, and then go trudging ahead without them. Illusions will still tug at us, most of us do not have the discipline of a Zen master emptying his mind, but even if we sometimes go where they tug, we give up, at least on the idea that these obsessions offer us any real protection against the big D. What then? What are we left with?

Well, nothing really. But on the other hand, everything, all that is solid—the world. Terrifying as that world can be once stripped of illusion.

My father was an ambitious man and ambition lured him south. A chance to run his own company, a larger company, a chance to be king of a wider realm. I, like my father before me, have pulled up stakes and landed in North Carolina. Like him, I am ambitious, though not for my teaching job. My ambitions focus more on my continued attempts to put words on the page, though I certainly no longer believe those words will live forever. While ambition still grips me, I have come more and more to recognize my ambition as phantasmagoric, no more substantial than the flecks of ash rising from a fire. In fact, it is the same airy stuff that "the feeling" was made of. My father's instinct, in the midst of his ten-year-old son's panic attack, was to rap his knuckles on something solid, like the living room table. This wood, he seemed to want to say, is the world, not those insubstantial ideas in your head.

And the world has *things*, like turkey vultures, in it.

—

There are flood and drouth
Over the eyes and in the mouth,
Dead water and dead sand
Contending for the upper hand.
The parched eviscerate soil
Gapes at the vanity of toil,
Laughs without mirth.
This is the death of earth.

—

Vultures are not a particularly popular bird for a variety of reasons. There's the death-eating, of course. But there's also the fact of their featherless faces. If we can impart some nobility to their soaring, we recoil when it's time for their close-up. Pink and undeniably penile, their heads reveal them as the cartoon villains of the bird world. Watching them through my binoculars, I can almost understand why people try to drive them out of their roosts, and out of some cities entirely.

But vultures have their defenders, too. One of those is the writer and birder Katie Fallon, whose book *Vulture: The Private Life of an Unloved Bird* is a deep look at the secret world of vultures. Katie is a good friend, and I would occasionally text her vulture questions from the deck in Boulder. Sometimes, when especially stumped, I would call.

One time I asked her if she thought there really was a pecking order in the tree I watched. There was no scientific consensus on this, she explained, but in the roosts she observed she had noticed that the young birds, distinguishable by their gray heads, hung out in the lower branches, and were therefore "pooped on" by the older birds above, which suggested that higher spots were more favorable.

She also told me that while turkey vultures may not like to be shat upon by other turkey vultures, they sometimes willingly shit on themselves. Katie described the way they let their own excrement dribble down their legs, not from mere sloppiness, but as protection: it coats and disinfects them as they walk among the dead.

"It's called urohidrosis," she said. "Think of it like vulture hand sanitizer."

I asked her about my summer's twin themes of emptiness and death.

"Sometimes vultures cause trouble in the mornings while they wait for the thermals to heat up. There were some black vultures who would hang out at the elementary school near my house waiting for the heat. They pulled shingles off the roof and vandalized the playground equipment. I think they were just bored."

Katie has worked closely with other raptors and in her opinion vultures are more intelligent.

"They just look at you differently than owls and hawks."

As for death, it's a concept that becomes a lot less abstract when you work with vultures all day. As part of her job, Katie locks herself in an enclosure with the birds for hours and she has wondered what would happen if she had an aneurism. How long would the birds wait? Those beaks, she said, could easily tear open a stomach.

—

The latest prediction for Hadley's future has arrived. It ain't pretty. Vultures may be thriving, but many other animals are not.

"2063 could see us well into the Sixth Mass Extinction,"

Anthony Barnosky writes me. Barnosky is a Professor Emeritus in the Department of Integrative Biology at the University of California, Berkeley. This world, he tells me, could well be one "without iconic animals such as elephants, rhinos, mountain gorillas, indeed without many of the species we now take for granted. That's if we continue business as usual."

As is common now in the doom business, he holds out the blessed if.

"If we actually get serious about addressing the current biodiversity crisis by ensuring that the little habitat remaining for wild species is not further diminished, and curbing over-exploitation of species, environmental pollution, and climate change, we may have a chance of avoiding that depressing outcome."

A chance.

—

William James wrote of two schools of religious thought, "the healthy-minded" and "sick-minded." The sick-minded see death as the fundamental reality of life and believe in the futility of all earthly endeavors, and therefore know humans to be limited creatures with distinct expiration dates. The healthy-minded see the joy of this life now, and think that human beings are creatures of unlimited potential.

Boulder was built with the bricks of healthy-mindedness. Its gods are Whitman and Emerson as well as assorted contemporary New Agers. Fitness and health—both physical and spiritual—are celebrated here as they are in few other places on the planet. It is easy to make fun of this, and I have, but there is also something stimulating and life-affirming about it. I couldn't have picked a better place to recover from cancer. The anti-Worcester. A kind of paradise. It is here that I began to bike every day. I liked the way that climbing hard drove out thought and doubt. It left me feeling indestructible.

If Boulder had a city anthem then surely the phrase "Being in the present moment" would be in it, but it's a phrase I've always liked to mock. Cows, I have pointed out to friends, are no doubt quite good at being in the present. Humans not so much.

To the sick-minded, the fact that we will die pulls the rug out from under any celebration of our momentary health or well-being. As James writes: "And so with most of us: a little cooling down of animal excitability and instinct, a little loss of animal toughness, a little irritable weakness and descent of the pain threshold, will bring the worm at the core of all our usual springs of delight into full view, and turn us into melancholy metaphysicians." And, "Let sanguine healthy-mindedness do its best with its strange power of living in the moment and ignoring and forgetting, still the evil background is really there to be thought of, and the skull will grin at the banquet." We may derive excitement from our work and plans and dreams but: "Let it be known to lead nowhere, and however agreeable it may be in its immediacy, its glow and gilding vanish."

—

Water and fire succeed
The town, the pasture and the weed.
Water and fire deride
The sacrifice that we denied.
Water and fire shall rot
The marred foundations we forgot,
Of sanctuary and choir.
This is the death of water and fire.

—

Near dusk they fill the skies above Boulder. Descending in spirals. It is beautiful. They rise again in the morning.

Whether one considers Boulder a human paradise or a smug, snooty, insulated prep school of a town (or perhaps both) is a matter of taste. But that it is a vulture paradise is undeniable. A city so close to the mountains is ideal. A direct rise to where the food waits. A direct descent to roost.

As you watch vultures you will notice that they do not do

too much flapping. This is because, while they have almost eagle-width wingspans, they weigh less than a third of what eagles do, and are not particularly strong. They rise by lifting on updrafts. There are two basic sorts of updrafts, thermal updrafts and mountain updrafts, and Boulder provides both in spades. A thermal updraft, or thermal for short, occurs when heat rises off the earth, and the vultures use these for lift. This is solar energy at its most basic and effective, and the best places for it are generally open, like fields and parking lots. You can't find solar energy rising off of trees, because the plants have already used that energy. Cities are perfect, virtual heat islands with their paved streets, with heat rising off the land like steam from a pot. Which makes a city next to the mountains a vulture dream.

The second kind of updraft occurs along mountain ridges, like the foothills above Boulder. In the simplest terms, mountain updrafts occur when horizontally moving winds hit the vertical sloping mountains and are deflected upward, like a ski jump. That is the jump the vultures take. Which means they can rise with little energy expended and with their heart rates barely more elevated than when they sleep.

So much for what they do for a living. But what of their internal lives? What goes on inside? Though a slew of books in recent decades have insisted that animals are smarter and more emotionally complicated than we have admitted in the past, most people are still loath to grant animals the complexity of an internal life. The main problem is lack of interspecies communication; they don't talk, write, or text so we don't know how, or how much, they feel. There are hints, of course. We watch the listless tiger pacing in its cage and we think "bored."

To Ernest Becker, humans are partly farcical creatures, animals who have grand plans but who drool, fart, bleed, and die. We assume no vulture has ever had a moment of self-awareness, has ever thought anything along the lines of "how strange that I eat dead things for a living." But watching their tree in the evening, you see clear signs of what most would call a very human restlessness. They assume a hundred postures, opening their wings to dry them in the wind, often shifting and shaking and turning, and they bully neighbors off their roosts, not for any great advantage, but, to my eyes at least, to pass the time.

How they fill their days may well be the result of evolution's programming, but are some days better than others? I have to believe that sitting in the roosting tree when fall's cool has come sweeping down the mountains is nicer than a stagnant summer day, and that the day you find the unguarded cougar stash in the pines—half a mule deer carcass!—is more vivid and engrossing, more memorable, than the humdrum scraping of the squashed squirrel off of Flagstaff Road.

I don't want to impute a more complex state of being to these birds than that which is merited. But do *Homo sapiens* feel bored solely because of the self-consciousness that their big brains bring them? Maybe boredom itself, the physical sensation of it, not just the notion of vacuity, is something that evolved in our earlier animal selves, and therefore, it stands to reason, evolved in other animals too.

I am not suggesting that a covey of dark-winged existential philosophers are circling above the town and pondering the fact of their own emptiness. Self-consciousness is not exclusively a human quality, ask any chimp with a mirror, but no one is denying that we are still the kings of that self-reflexive domain.

One defining characteristic of humans is the sheer variability of behavior. Which is a big part of our problem. The fact that it is choice, not just evolutionary necessity, that determines how we fill our days. We are not just faced with an emptiness and overwhelmed with possibilities, but are held responsible, in the court of our own psyches, for the choices we do make.

—

Writing about climate over the last two decades has been frustrating. I think this is in part because, as a writer, my goal is to show what is, not what I hope will be. I am not a politician. It is not my nature to write to convert, to argue. Rather I try to bear witness. And yet…

And yet how to make people see? What kind of language will work? Will any? How to make people act? And, to complicate things, is that really the goal of literature? Has it ever been?

Maybe it's just not very grown-up to think that there is an *answer* to the chaos, or that we, and more importantly our governments, having done so little for so long, are going to suddenly change our behavior. As the pandemic and the vaccine make clear, logic doesn't always hold sway. Nor does common sense. Common sense would tell us that global warming, not the other issues that are pushing upon us at the moment, is *the* pressing issue in the world. On a planet that is burning and flooding, where life is inhospitable, and fewer and fewer places are livable, climate is not just one of many possible dramas being acted out. It is rather the stage on which *all* the dramas are performed, and we are losing that stage. And at this squabbling point on this over-connected planet, who can be confident that we will respond appropriately, even with our lives at stake? As my friend the physicist Dylan McNamara succinctly puts it: "We're fucked."

But even if we *are* fucked, it is worth it, I believe, to, as we like to tell preschoolers, *use our words.* And to use those words to honestly lay out the situation we are in. Not the worst-case situation or the best-case in the manner of some of the peppy, techno fixers. The world *as it is now.*

—

My father feared death his whole life. But when it actually came he responded well.

I was with him during his last days. His death took over the book I was writing at that time. I typed and scribbled straight through those final days, taking few breaks except to care for him and write his obituary. My father was only fifty-seven when he died, four years younger than I am now. Two years after his death I faced another. I hadn't gotten close, really close, to a dog since Macker, but as I was falling in love with my wife, I also fell for her dog, Zeke. Zeke was a curmudgeon who bit many and loved few. Part Saint Bernard, part collie, he took his last watery breaths in the garden behind our house. Zeke's death has stayed with me, in no small part because of how closely it resembled the death of his fellow curmudgeon, my father. Both animals breathed long, labored breaths as they closed in on death, then shallower gasps at wider intervals as the moment approached. Both, Zeke in the unplanted mud of the garden and my father in his hospice bed, seemed ready to let go before hanging on and fighting back, this pattern repeating over and over. Their eyes looked off behind them and away from us, but they were clearly aware that we were holding onto them. The sheer physicality of the moment was like none other, the only thing comparable for me being the final moments of my daughter's birth.

Another way that the two deaths were alike: neither took solace in religion.

My father was a nonbeliever until the end. Did this soothe him? Doubtful.

But after knowing him my whole life I would have been shocked and disappointed if he made that cowardly retreat to God in his last hours.

Don't get me wrong. I am full of admiration for the man of faith who stays faithful in the face of this sternest test. That is every bit as admirable as my father's death. But what is not so admirable is, having lived one way, trying to suddenly fudge things as the end nears. "I'm sticking to my guns," my father said in his usual businessman vernacular. And he did stick to his guns.

I, like my father, am a nonbeliever. So what does that leave me with?

Quite a lot, it turns out. For one, this world—its smells, its tastes, its feel, and of course its people and birds and other animals. And if I have learned one thing in the years since "the feeling," it is that the world exists quite separate from me, thank you very much. But what it doesn't leave me with, in the end, is the self. When I die I am gone. *Kaput. Poooof.* I can't imagine that I will turn to prayer. As for ambition, what will that get me? Perhaps a slightly longer obituary.

But this is too glib an answer. Even if my work is not remembered, it gives me much. Not the least of its gifts is that it fills me up while I am here. But it is more than that. In the end my work is the something I make out of the nothingness, even if it does not help me escape oblivion. It is my sacrifice made at an empty altar.

And what of the world, which, I hope, will remain when I am gone?

It is too late for me to alter my own custom-built blinders, at least by very much. I am not built to save the world. But I'm pretty good with words. That is my skill and those are my tools. This is what I can bring to the big table. I'm not dismissing or belittling this—words are fairly important.

But this dark time coming toward us is going to require more than words. There are others with other skills who are going to have to fine-tune their purposes to fight it. Some have the tools of science. Some the tools of inspiration. Some of administration. And some, whom so much depends on, the tools of politics. We will need people with well-built blinders. And we'll need all hands on deck, lives of purpose focused on healing what we've done to our cracked planet. We'll likely fail at this point, but that doesn't slow you down if you have that purpose thing. It seems a worthy enough cause, a good enough purpose, after all. Saving the world.

—

If we are going to give pep talks, perhaps they should be laced with realism.

How about this one from Mr. Eliot?

For us there is only the trying.

—

Here is something Katie Fallon said about vultures:

"They are a nagging reminder that something is waiting for us all to die. That something is waiting to consume our bodies. Something whose survival depends on whether we die or not. They remind us that we are animals. And that like every other animal we will be eaten by other animals."

She added: "It's one thing if we are eaten by worms or bugs. But a vulture presents a more in-your-face sort of consumption."

Throughout our summers in Boulder I have biked up the hills every day while the vultures go about their daily work of ascending and descending. It might seem to us that picking apart and digesting dead things and shitting on one's own legs is not the most pleasant way to pass a day. But at least there is pleasure in knowing exactly what one's daily task is. As humans we are not assigned our lots but are faced with choice, though we may have less of that than we think. But even if real choice doesn't exist, it certainly seems, to paraphrase James, as if it does. And with choice comes regret and indecision.

Which, we think, the turkey vultures are blissfully free of. And that must be, we may conclude as they soar up into the mountains in search of death, a kind of paradise on earth. Meanwhile I get up each morning and join the rest of Boulder in its desperate festival of fitness, hoping to hone my body while short-circuiting parts of my brain. Unable to gracefully ride the thermals, I labor at my more willful ascension, pushing down on the pedals of my bike, sweating and grunting with animal determination, and occasional animal pleasure, focused on the Sisyphean task of getting up the hill while trying not to notice the dark birds circling overhead.

PART III.

HURRICANE SEASON

"We need more hypocrites who fight."

—Dan Driscoll

TRAVELS WITH ORRIN

Back home in my adopted state of North Carolina, after a summer out West, I wait. Or rather we, as a family, a city, a state, a region, wait. August—*prime time!*—rolls around. Then September, even more dangerous. When will one come? Will it be the Big One? The angry drunk? Will our lives be overturned?

We live in uncertainty. Spin the wheel and see where the arrow points this year. The Carolinas? Southern Florida? New Orleans? The Panhandle? All of the above?

We'll see.

—

Isabel was our first. We had barely settled, barely unpacked really, when we heard the word that the Big One was heading toward us. We were the bull's-eye—the dashing Weather Channel reporter had set up shop at the pier near our house, using it as a stage from which he could narrate the storm. Hadley was just four months old and we had no idea what to do. Then the word changed a day or so before the storm landed: the Big One would hit elsewhere. The Weather Channel reporter charged off to another pier farther north.

But if Isabel did not hit us directly, we still felt its lash. On that day—September 18, 2003—I decided, perhaps foolishly, that I wanted to see the storm from the shore, near the new apartment we had rented on Wrightsville Beach. So after I'd tucked Nina and Hadley safely in a Holiday Inn in town, I drove back out to the island, ostensibly to check in on how our rental apartment was doing. I am not a thrill seeker, or at least not an extreme one, but I like to watch big storms come in over the water. There is pleasure in all that force, until it crosses a certain line.

I would soon learn that the surfers out at Wrightsville Beach keep riding the great swells to the very last minute, trying not to quit until that perfect moment when wild delight becomes true danger. The philosopher William James, who happened to be in San Francisco for the earthquake and fire of 1906, would later write of the devastation to the city, the horror, the tragedy. But he also mentioned the "wild Olympian joy" he felt right after surviving the event itself.

If my joy was not quite Olympian that day, I could feel something rising. Driving in, I stared up at phone lines bobbing wildly and street signs looking and sounding like they were being slapped around. Water spilled over the causeway. With the surfers and most of the weather reporters finally gone, I had the beach, and the hurricane, to myself. Sand lashed my back, and water too, the rain falling almost parallel to the ground.

After a while I walked inland, toward the houses, and noticed that there was a human energy that almost matched nature's. The few people who were still outside moved with purpose, intent on their last-second, and last-ditch, efforts to board up, tie down, and get out. I remembered the feeling from childhood, when I spent summers next to a harbor on Cape Cod, the hurricane fear but also the hurricane excitement, the sense of a community readying, a community in danger, but a community that was never more a community than at that moment, as neighbor helped neighbor prepare for the storm. And I remembered my little sister, barely six, jumping in the air off a dune with her windbreaker outstretched like wings before the storm hit fully. There were some jumps when she got enough lift that I swore she was flying.

Before we moved to North Carolina in 2003 it didn't occur to me that I should be nervous about hurricanes. Nor did I quite understand that we were moving to the southern coast right in the middle of hurricane season. We would soon get used to the feeling of growing anxious every summer, as the calendar turned to July and the ocean waters heated up. It is a feeling, one so common down here that we barely mention it to each other, that those in the northeastern part of the country will soon have to get used to.

At that point in my life I had experienced a couple of hurricanes, but they were northern hurricanes I see now, not *real* hurricanes. There was even an element of fun to them that you weren't supposed to mention, an excitement that you couldn't be entirely honest about. It is something I've since noticed down here too, though our southern storms are much more deadly. You are supposed to wear an expression of fear and concern, but it is not uncommon to see goofy smiles and wild eyes.

This is one of the problems with hurricanes, or of our perception of them. Yes, there are deaths and damage, but few acknowledge just how wild the experience of being in one is.

Of course there is little joy when your home is destroyed, your world left in shambles. But either way, hurricanes reshuffle our decks. Here it comes and with it your whole life can be changed, can be reordered. Not through your own feeble efforts at reform, your daily self-injunctions to "improve," but through a force much larger than you and your petty concerns, a force that momentarily pulls off your veil, snaps off your blinders. You may usually concern yourself with work, or even with ideas, with concepts and notions out of which your priorities grow, and perhaps you even have a kind of habitual mental system by which you rank, control, and justify the unruliness of your actual life. But this thing that is coming, this thing about to change your life, is not an idea. It is—get ready for the punchline—*wind*.

Just think of that. Wind, the mere movement of the earth's air. Could there be anything less relevant to our cherished notions of what constitutes our neat, modern, supposedly connected lives, less relevant to our cherished goals and plans and efforts? But as it rises off the water and hurtles toward you it is about to show you that it is not irrelevant, far from it. Coming into town like a great gusting bully, rattling those street signs and plucking telephone wires like giant guitar strings, it will show you that it is in fact very, very relevant; it is about to blow your little piggy house down, no matter how many bricks you pile up.

It comes charging in, bursting in, blowing in, gushing in, storming in. And it brings with it building excitement and, of course, fear. The ocean spits forth foam as if rabid, and then builds up with great humpbacked power. People rush this way and that, as if boarding up some windows will stop this wild thing. But they have to do something, you see, anything, to reassert some small degree of control.

—

Perhaps it was my own attempt to gain some control, to find some answers, that led me to Orrin Pilkey. During my first visit with Orrin, in his book-cluttered office at Duke, he wasted no time laying out what he called "the new math" of living by the shore. This was 2006 and he explained how over 153 million people lived at the coast in the United States, an increase of 33 million people since 1980. (This number varies depending on what you call the coast, though generally it means that 40 percent of the population lives on 10 percent of our land mass.) How these people were building larger and larger homes closer and closer to the sea just as the shoreline was eroding and the seas rising, not to mention the fact that storms in our overheated oceans were becoming more violent. Long before others had, Orrin saw this combination of forces as the recipe for disaster it is. He spoke of sea level rise and the encroaching sea and the overbuilding and overpopulating of our shores. He connected the dots.

That day in his office I learned that Orrin was born in Brooklyn but grew up in Washington State, far from the sea. His first memory of the coast was of feeding the pigeons on—irony of ironies—the place that would become the subject of his wrath: the Jersey Shore.

Not only had Orrin not grown up on the coast, he didn't live there now. Teaching at Duke, he was over two hundred miles from the ocean, and I mentioned the fact that he—the great defender of the shore, the famous reader of beaches—lived that far inland.

He made no apologies.

"It's probably best that I don't live too close to the beach," he said. "Considering my opinions."

I saw his point. People wouldn't be too crazy about having a neighbor telling them that they shouldn't fight to defend their homes, that they should simply give them up to the sea. I imagined broken windows, nasty notes, other threats and snubs.

The fact of where we each lived pointed to an important difference between Orrin Pilkey and me. For Orrin the question of how to live near the water was a professional one. For me it was personal. Later, when we got to know each other better, I would come to understand that he actually got a kick out of needling us beach-lovers. When we traveled I sometimes took off my shoes and walked barefoot on the beach. But just as I started to enjoy the feel of sand between my toes, he would remind me of the high fecal content on beaches due to all the dog-walking.

In fact you could say that Orrin came to beaches not through sand and sea, not through his body, but through an idea. It was not his own idea either, though he certainly had his own reaction to it and he would soon enough make it his own.

The idea was handed to him when he was thirty years old and out to sea.

"I was a deep sea sedimentologist," he told me. "I didn't know anything about any of this beach stuff. But I was out on the research vessel *Eastward* taking water samples with a scientist named Jack Pierce, who worked for the Smithsonian, and since neither of us were the principal investigators we had a lot of down time while out at sea. We played cards and talked. Jack talked at great length about how all these people were building on an eroding shoreline and then expecting the government to bail them out when their houses were destroyed by storms. And he talked about the inefficiency of the Army Corps of Engineers. This was the late sixties when no one knew about this stuff, but Jack did. He talked about the fact that groins and armaments were ruining the beaches. At the time everyone was building seawalls. He told me about what they'd done in Jersey and how it had ruined the beaches. The funny thing is that this stuff didn't really bother him. It didn't make him angry. He was just curious about it and well-informed. For him it was kind of fascinating. But when he told me, I had a different reaction. I became incensed. Outraged, really. That's how I got started."

Orrin was already teaching at Duke, but meeting Jack Pierce would change the course of his career. When he got back home, he started doing some research and found out that his colleague on the ship had been right.

"Early on I learned that New Jersey was really the centerpiece of bad coastal management," he said. "Some of our first beach resorts were in New Jersey and they began building walls to protect their beaches a hundred and fifty years ago. Well, who wouldn't want to build a seawall to protect a beach? A hundred and fifty years ago who wouldn't say that is a good thing to do? Of course you can't let the houses fall into the ocean."

But what New Jersey proved, he learned, was that seawalls, groins, and jetties ultimately damaged the beach. In fact, *damaged* was a mild term. The beach would keep retreating toward the fixed object until it was swallowed up entirely. You were left with a wall with water sloshing against it. No sand and no slope.

"After studying seawalls for a while I began to see that they destroyed beaches. I began advocating against building them, saying we should retreat from the coast and let buildings fall into the sea. Of course the Corps of Engineers, who built many of the walls, didn't like the word *retreat*. They thought it quite unmanly.

"I now see New Jersey was a giant science experiment. Thanks to New Jersey we would learn that any sort of hard stabilization—seawalls, groins, and jetties—was very damaging to the beach. We learned that the damage occurs just by building something fixed by the beach—could be a highway, for instance. The problem with beaches is that they are eroding and always moving. The beach tends to move toward that fixed thing and get narrower and narrower and narrower until it disappears altogether."

The result of Orrin's research was a book called *How to Live with an Island* that he cowrote with a scientist named Robb Turner and with his father, Orrin H. Pilkey Sr. The book came out in 1977 and was his earliest attempt to express his philosophy of retreating from the beaches. In the wake of that book, Orrin got a lot of responses from people. And that was when he figured out that he *liked* getting responses. Even the angry ones.

"That's how I learned that I really like stirring things up," he said.

Thirty-nine more books would follow. In them his mind ranged far and wide but always came back to the central idea that the way we were living on the coast was the wrong way.

During that first visit with Orrin I noticed that he had trouble looking from side to side. Then he noticed me noticing and explained.

"It happened over thirty years ago," he said. "My students and I were doing research up on Shackleford Banks. I climbed up on top of the mast to dive off. Showing off, of course. But the boat had swung around and the water was shallower than I'd calculated. I dove off the mast and broke my neck. Would have died right there in the water if my favorite grad student hadn't grabbed me by the hair and swam me back to shore. They took me in an ambulance all the way back to Duke."

This led to more stories of traveling the Outer Banks with his students, and before long he was feeling nostalgic and suggesting that we take a trip along the North Carolina coastline together.

"I used to do it every year with my classes," he said. "I'd love to explore the coast with you. Show you some things."

And so our travels began.

—

Coffee proved our initial bond. As it turned out, Orrin had recently heard a radio segment on McDonald's coffee, on how they had upgraded what they served, and after I picked him up in Durham he suggested we see for ourselves. I pulled into the first drive-through we saw and both of us bought large cups. I would soon learn that my traveling partner relied on these steady infusions of caffeine, amounts that made me jittery but that in his case merely maintained a high and steady level of energy.

We drove on through the flatlands of Carolina toward the ocean. We were still a hundred miles from the shore, but the land looked sunken to the point of inverse, a great bowl of a place filled with pond-sized puddles from recent rains.

"People worry about the beach and the cities because that's where the money is," Orrin said. "But they should also worry about places like this."

He was referring to the coastal plains, plains that in North Carolina extend back to Raleigh, and that in Georgia extend almost all the way inland to Atlanta. Orrin explained that in places like these, unlike places on the often mountainous Pacific coast, a small sea level rise could spread far inland. The land is low and the rise only slight as it moves away from the ocean.

"If the seas rise seven feet this is all underwater too," he said.

This was before people began using the phrase "ghost forests" to describe the trees killed by the inundation of salt water. As for exactly how high the water would rise here, on these plains, relative to the overall sea level rise, that is open to some debate. If the beach were perfectly uniform and had no irregularities, the water would move inland at a rate of about two thousand feet per one foot of sea level rise. This is because the slope of the coastal plain here is roughly one to two thousand.

Of course there are no perfectly uniform beaches, and irregularities are what coasts are all about. Personally, Orrin had little use for theoretical beaches, for what he called "beaches in an expected universe." He was particularly disdainful of the Bruun Rule, the formula that engineers had long used to calculate the way that water inundates a coast.

"The Bruun Rule has been discredited for years now, but some engineers still swear by it. They love it because it gives them a formula. But the composition of every beach is different, and the factors—wave energy, ocean floor composition, downdrift erosion—are different too. As much as the engineers would like to claim there is, there isn't a single rule that encompasses all the factors."

Orrin was hungry so we stopped at a mom-and-pop restaurant to gobble down plates of ham and eggs, then fuel up, as would become our tradition, on more coffee. On the bridge over to Manteo you could really see how flat the land was, how close to being water. We pulled off the road and drove down a dirt path into the Alligator River National Wildlife Refuge. These swamplands supported this country's only wild population of red wolves, the result of a successful recovery program. On a recent visit with my daughter, then seven and wolf-obsessed, we had seen not just wolves but a black bear peeking out of the woods. And yet the ecosystem they lived in was mere inches above sea level.

After Alligator River we crossed over the next bridge to the Outer Banks, the series of islands that jut out into the Atlantic. On the Outer Banks maps get outdated quickly. These islands and the watery spaces between them are always in motion, sometimes moving quite dramatically in a short time. During hurricanes new inlets can and do cut across the islands and straight through homes and roads, while existing inlets can open from a hundred yards wide to a mile in one day due to massive amounts of water pushed in, and then funneled and flushed out, by the storms.

This makes the building of bridges tricky. On the way down to the Hatteras, for instance, we would be crossing a temporary military-style bridge. Sometimes bridges will span what appears to the non-engineer to be dry land, an appearance due to the fact that it actually is. The movement of both water and land here is highly unpredictable.

If the predictions are real and we are indeed going under, then it is this place that is going under first, but that fact hasn't stopped rampant development, a kind of sprawling rich man's ghetto along the shore. Huge trophy homes on stilts stand packed close together like gang members.

"There are some respected geologists who doubt the Outer Banks will survive the century," he said.

He let out a deep sigh, as if the idea pained him.

"But I'm slightly more optimistic. People may not live here but the islands themselves could survive. As storms increase in intensity, barrier islands will serve as our front lines. The islands are fairly well built for this role. They are in some ways the most dynamic natural environment on earth. They maintain themselves as they move back. A storm will hit them and sand will flow over them and replenish their backside, which is usually marsh. In this way they are always on the move.

"People fear storms but barrier islands need storms to live. Storms feed the islands. They are the way the islands migrate and the way they build elevation. If sand is not pushed across an island by storms then the island drowns. The trouble comes when we start building roads and houses on these moving islands."

As we drove through the towns of Whalebone and Nags Head and Kill Devil Hills, we saw evidence of this trouble. First there were the usual clustered groups of McMansions, suburbia on the sand, in places where just a few shacks and no other homes had stood when Orrin first studied the area. Then there was the road itself, less than fifty yards from the ocean in many places. We stopped the car and Orrin pointed out how the sand grew up high on either side of the road so that we were driving through a manmade valley, dunes on both sides. He called the road an "overwash canyon," a geologic term, he proudly told me, of his own invention.

"You see the island is trying to do the right thing and we don't let it," he said. "It wants to move back from the water and gain altitude. Instead we just push the sand back off the road toward the water."

We passed motor lodges that time forgot, places with names like the Seahorse Motel or the Mariner, still lost in the fifties and sixties, squat buildings that barely came up to the knees of the new condos and mansions. Orrin pointed out a hotel where a cabin labeled "Number 3" stood closest to shore; he explained that numbers 1 and 2 had already floated out to sea. Most of the houses were elevated on stilts, as required by federal flood regulations, though some owners had illegally filled in their bottom levels, slapping walls on the stilts to form a ground floor.

Back in the seventies, Orrin and his fellow coastal scientists had watched the rampant development of these coasts and figured that once the first hurricane blew through people would know better than to build here. He quickly learned this was not so. Instead hurricanes became "giant urban renewal projects." Orrin called this cycle "societal madness," the way rebuilding begins almost immediately.

"They think they are rebuilding on the same spot but the spots have likely doubled in danger."

We climbed back in the car and crossed onto Hatteras Island, traveling through undeveloped land that was the nation's first national seashore. A high dune line ran for miles to our left, protecting the road from the ocean, and at first you might have thought it a natural one. But Orrin informed me that it was the result of massive human effort, a Depression work project funded by the Works Progress Administration and National Youth Administration that employed hundreds of men and boys.

"All to protect a road," he said, shaking his head.

When we reached the first outpost after the dunes Orrin had me pull over to show me the old cemetery, where dozens of the workers from the work project were buried. He pointed out that the more contemporary graves faced in one direction, toward the current road, but the older Depression-era graves faced another, where the old road was and where the sea is now.

Not long after, we pulled into the parking below the Hatteras Light. Deer nibbled at the bushes. The striped lighthouse rose above us like a giant barber pole. The Hatteras Light was and is a beloved landmark to North Carolinians, a 208-foot-tall black-and-white-striped brick tower that shines its light out over one of the most dangerous stretches of shoreline, the famous "graveyard of the Atlantic."

Perhaps even more than on Topsail Island, it was here on the Outer Banks that Orrin was most admired and most despised. It isn't just that he points out the foolishness of building on flat, eroding land so close to the ocean. It's what he says next that really bugs people. He tells them that by trying to defend their homes they are actually destroying the beach, which is what ultimately protects the shore. He tells them that their natural urge to defend themselves, to build a wall or pile up sandbags against the encroaching water, is wrong. That by doing this they are keeping the beach from doing what it wants to do, what it *needs* to do, which is naturally migrate up and over itself. LUST is the catchy geological acronym for this movement—landward and upward in space and time—and walls get in the way of LUST, deflecting the sand back. This movement is how barrier islands have always defended themselves from storms. In fact, *defend* is the wrong word, since storms help islands grow. What a wall does, on the other hand, is draw a line in the sand, cutting off the shore's natural slope and movement. This is what Orrin tells homeowners. It is, of course, exactly what they don't want to hear.

The locals haven't burned Orrin in effigy, at least not yet. But there have been threats. Especially when Orrin spearheaded the retreat of North Carolina's most famous landmark. Locals hated the idea of moving the lighthouse back from the shore, refusing to admit that if they didn't it would fall into the sea.

"There was one powerful local woman who was virulently opposed to moving it," Orrin told me as we walked below the lighthouse. "She said, 'Someone is going to get hurt if they move it.' A fellow scientist misunderstood and tried to reassure her. 'No, Mrs. Dillon, we can move it perfectly safely.' I had to explain to him that that was not what she meant."

Orrin and I walked the 2,900 feet from where the lighthouse had been built in 1870 to the spot where it had been moved in 1999.

"Mrs. Dillon always claimed that moving the lighthouse killed her husband. The stress, you know."

I asked if Mrs. Dillon had passed away too.

"She's still alive," Orrin said. "Unfortunately."

In fact, he told me, Mrs. Dillon had recently bragged about staying through a hurricane and managing to save her house by shutting an upstairs window.

He shook his head. "People think they can protect their houses, but what can you do when the winds are blowing a hundred miles an hour? It isn't brave to stay, it's stupid."

We reached the location where the old lighthouse once stood. When it was built in 1870 this spot had been 1,600 feet from the shore, but by World War II that distance was reduced to 500 feet. Today the stones that mark the old foundation are on the sandy beach itself.

"In 1972 there was a big storm and the waves came right up to the foundation stones," Orrin said. "In the middle of the storm the Park Service tore up the parking lot and threw the chunks in and saved the lighthouse. I began to argue that we needed to move the lighthouse back. Then about a year later they had a big meeting of all the players and I didn't get there and that was the biggest mistake I ever made. The meeting was taken over by a local booster who argued that if we moved the lighthouse we'd make fools of ourselves."

The booster won the day and the lighthouse stayed put, but the sea kept creeping closer. Groins were built and the beach was "re-nourished" by adding imported sand while sandbags were piled up at the base of the lighthouse itself, but still it was obvious that it was eventually going to fall. Orrin then formed a "Move the Lighthouse" committee. He enlisted the National Academy of Science to do a study that showed that the lighthouse could be moved inland. His committee published posters with the caption "Move It or Lose It" and he thundered this same message around the state, drumming up attention and support. The folks who opposed him—"the bad guys," he said with a laugh—were the usual beach town suspects. The businessmen, the boosters, the rah-rah realtors.

But it wasn't just the locals who fought the move of their beloved landmark. More prominent North Carolinians spoke up too, those who usually didn't get their hands dirty with mere local issues. Why did it resonate so? Because the Cape Hatteras Lighthouse *was* North Carolina, and to move it back meant you had to admit that other buildings—including the summer homes of the rich—might have to be moved back. The lighthouse had always been there, after all, and now here came the dreaded Pilkey, the great advocate of chaos and uncertainty, braying about how their beloved landmark was going to topple into the ocean. It was *appalling*, really.

To move the lighthouse was also to admit that nature had the upper hand, to accept succumbing to something wild beyond your control, and though many pay this idea lip service, few who live at the shore really want to accept it. But finally the Park Service, pressured by the North Carolina Division of Coastal Management, capitulated and agreed that the lighthouse would have to be moved. The move was a spectacle and Orrin was right in the center of it, the ringmaster, even taking a turn pulling the lever that "drove" the lighthouse along the tracks that had been laid to transport it to its new home a half mile inland and upland. There had been some applause when he took the controls, and, no doubt, more than a smattering of boos and heckling.

Before we left I played devil's advocate and pointed out to Orrin that the spot where the foundation stood was not yet underwater.

"That's true, but look at the lengths they've gone to protect it."

He pointed to the three metal groins that jutted out into the water and then at the downdrift beach, which had been carved away by what the groins had done, eroded in a scallop shape.

"The point I'm trying to make is not that *I* can predict what is going to happen. It's that no one can. For instance, right when I was arguing for moving the lighthouse, a hurricane hit that actually ended up *adding* sand to this beach. Who knew?"

It occurred to me that Orrin's logic also applied to the houses built in the foothills and mountains in the West. It is commonplace now that those with the least will be hurt the most by climate change, and on a deeper, more profound global level this is obviously and tragically true. But scientific studies show that in the United States the wealthy, with their homes making incursions into rural places that until recently were allowed to flood or burn naturally, have lost far more homes than the poor. Anyone driving along the Atlantic beaches or foothills of the Rockies could reach the same conclusion. Million-dollar homes. Kindling and driftwood.

From the Hatteras Light we drove back north through the towns of Buxton, Salvo, and Rodanthe. Buxton was where the Outer Banks' love-hate relationship with Orrin began. In the seventies, the Park Service, after paying heed to the geological studies, determined to stop artificially building up dunes around the town. Soon after that, Orrin had come to town with a reporter and announced that the Park Service's new policy was "to let Buxton fall into the sea." That was when he became public enemy number one.

We headed to the Comfort Inn in Nags Head, a choice of lodging both apt and ironic since it epitomized the exact sort of edifice that, in Orrin's view, should never be built along the shore.

"The worst thing you can do on a shoreline is build something that can't be moved," he explained. "It takes away all flexibility. That's why I worry that Florida is due for a truly epic disaster. What can you do when you have hundreds of miles of high-rises on your shoreline?"

Orrin had written extensively about the hotel where we were staying, and before we checked in he worried about being recognized at the desk. He told me that he used to stop here on his annual trip with his Duke geology class and that one year they had had the good fortune to arrive on the night the swimming pool caved in, its shallow end hanging down over the scarp onto the beach like an open jaw.

Orrin explained to me that over the years the hotel's name had changed from Ramada to Armada to Comfort, but that the one constant has been its role as a kind of symbol of why you didn't build any sort of high-rise along the coast.

"The problem with high-rises like this is you can't move them," he said. "And if you can't move them the sea will eventually drag them down."

After we checked in, we walked around back and I saw that the ocean was well on its way to dragging down this particular building. The Comfort Inn cantilevered out over the water. This was due, not to some sort of Frank Lloyd Wright fit of inspiration on its builders' part, but to nature's gnawing away the beach below it. Hundreds of sandbags leaned against the stilts that supported the building's base. But to even call them "bags" is to not get the point across. They were enormous, ten feet long and terrifically ugly, great lumpish loaves that transformed the beach into a war zone.

Orrin walked over and kicked one of the sandbags and pointed to where it had been ripped open. He speculated that it could have been from a sharp object—a piece of wood with a nail?—thrown up during a storm, but then he wondered if it might have been eco-sabotage.

"Do these look like knife cuts?" he asked.

He said this with a gleam in his eye and something close to a pyromaniac's excitement.

The sandbags gave the place the look of a war-ravaged place, a massive bastion against the sea. They brought to mind pictures of trench warfare in World War I. While long-term use of sandbags was officially illegal in North Carolina, these bags had been allowed through an emergency exemption that had apparently lasted a period of years. At ten feet long and seven hundred pounds heavy, they would not be easy to get rid of, no matter what you called them or what the laws were. At any rate their removal would require the same heavy equipment that helped get them here in the first place, and I wondered who was likely to pay for that equipment when the goal was taking them away.

The problem with piling sandbags is the same as with seawalls and jetties and groins. While they might temporarily protect a particular building, they cause downdrift erosion. This means that the next house down, the house to the south, bears the brunt of the erosion.

Orrin said he was going to head up to rest before dinner, but he suggested I take a walk south to see more stranded homes. Sure enough I soon came upon another row of houses that stood directly on the low-tide beach, hundreds of sandbags humped in front of them. While piling up sandbags might understandably be viewed as self-protective—*I can't let my house topple into the sea*—it is also, at the very least, an unknowing assault on one's southern neighbor. Because if you put up sandbags, the next house down the beach will have to put them up and so on and so forth. For a long stretch every house I passed was fronted by the great lumpish bags, as if hundreds of small, drab-colored whales had decided to beach themselves all at once.

We had requested the rooms at the far end of the hotel, which hung out over the Atlantic, where you could feel the waves in your sleep. But before I went to bed I walked out on the balcony and listened to the surf crash into the building below. I wondered: What would happen if they really removed all the sandbags and walls along this shore? The buildings would surely go, at least some of them, during the next big storm. I indulged in a little daydream, picturing Orrin standing down on the beach, staring up as the Comfort Inn—formerly the Armada and before that the Ramada—buckled to its stilted knees and fell forward into the ocean. In my vision I armed Orrin with a boom box on which he cranked up Beethoven's Ninth. A slight smile formed below his beard and a glint lit his eyes as concrete crashed into the surf.

—

We woke the next morning to a blazing red and orange sunrise over the ocean. I took a morning walk on the beach up to the pier that my friend Chip Hemingway had designed, its wind turbines whirring in the off-shore breeze. The building pleased me, with its turbines and cedar shingles, and even made me feel a little hopeful for the place. It didn't arrogantly jut in the manner of some piers, but pointed instead toward the future of the Outer Banks, assuming that that future is not underwater.

When I got back, we packed the car and prepared to head to points north, but we couldn't resist the draw of a free breakfast. Once that had been consumed, we climbed back in the car and headed up toward the town of Corolla. Corolla, like Ocracoke, is famous for its wild horses, horses that roam freely in the dunes. But a gate that ran all the way down to the water had been built to keep them from wandering farther north. Another reason the gate had been erected is that the beach to the north has been converted into a four-wheeler's playland. If you kept driving north along the beach you would cross the state border on sand, an entrance from Virginia into North Carolina (and vice versa) that doesn't appear on any road map. Orrin said that until another fence was put up near the border some people commuted to work between the two states along the sand.

We had come to Corolla to visit the Audubon Sanctuary, which would soon be redesigned by Chip Hemingway, the same friend who designed the Nags Head pier. Chip had given us the number of Mark Buckler, the Audubon Sanctuary's director, and suggested we pay a visit to the site of his new project.

He had also told me that Mark was a fan of both Orrin and my writing, which led to some lively speculation, between Orrin and me, on whose books he liked better.

It turned out that Mark, with the wisdom of Solomon, had copies of books for both of us to sign.

He also had some interesting answers when Orrin asked him how they planned on protecting the sanctuary's land from the rising sea. The plan was to attempt to fight erosion not with walls but with oyster beds and other "soft" defenses. Hard defenses, like seawalls and terminal groins, destroy not just beaches but the ecosystems of fish, birds, and crabs.

"That's fascinating," Orrin said as we drove away. "That's really the way to do it. Don't fight nature. Let it work *for* you."

But then his expression turned sour.

"That's exactly the opposite of the way the Corps thinks about the shore," he said.

The Corps, I knew, were the Corps of Engineers, the one group most responsible, other than private homeowners, for the arming of the beaches. It occurred to me that if you considered

Orrin a kind of superhero (with a big red R for Retreat on his chest) then his arch enemy was the Corps of Engineers, a group that, like crazed beavers, was always erecting hard barriers along the coast, no matter how the scientists protested.

"I think engineers are the major threat to American beaches," he said.

"You really hate those guys," I said.

"Well, if you ask an engineer to solve a beach problem, you can't be surprised when he gives you an engineering solution."

The basic problem with this is that it rules out any philosophical debate over whether a beach should be armed or not. It starts with the premise that houses must be protected, no matter the consequences to the ecosystem, and therefore sand is piled, barriers are erected, walls are built.

"Believe it or not I've actually come to sympathize with them a little," Orrin said. "Because I've now come to think—very, very clearly—that the Corps' hand is forced by Congress. Congress has kept them beholden to them. The congressmen need projects for their districts and the Corps are the product of the Congress. They do their bidding, and are forced to do ridiculous things to survive. The Corps would change if the Congress really wanted an honest accounting. They don't. They want projects for their districts. Some of them have been in the game so long that they think it's okay to lie. Because that canal we need for our district is very, very important. Sure, we know our facts are wrong, but it's still important."

Thanks to Orrin and his colleagues like Stan Riggs at East Carolina University, North Carolina had been one of the most progressive states when it came to coastal management. But that was changing quickly. State legislators had recently made themselves into national laughingstocks when they tried to legislate against sea level rise. This began when a group of respected scientists handed in a state-commissioned report that suggested that it would be prudent to anticipate a one-meter sea level rise along the state's coastline by the year 2100. Not so fast, said a group of coastal developers, imagining all the soon-to-be-underwater land they could no longer sell. With Orwellian brilliance, the developers decided to push for a ban—not on actual sea level rise itself (which was, even they conceded, impossible), but on any language that admitted to it. The legislators happily agreed. And so Orrin and his ilk were now cast as the state's Galileos, its most forceful proponents of the science deemed heresy by the nonscientists. Orrin said that not long after the ruling, the government seized his friend Rob Young's computer for mentioning sea level rise.

The truth is that, despite ever-more-sophisticated computer models, no one really knows how much the sea will rise. Computer models are in fact a pet peeve of Orrin's: he believes that scientists, after years of training and research, should trust their intuition and instincts, in the manner of artists, more than they trust computer models, and he has written a book about this concept called *Useless Arithmetic*. The thorniest problem in modeling sea level rise is anticipating the behavior of the glaciers.

Even Orrin, while using seven feet as his "working figure," is skeptical of anyone who speaks with too much certainty on the issue.

But in today's twisted political climate, where mainstream scientists, once revered, are now routinely doubted by nonscientists, it does not pay to voice too much uncertainty. That's where Orrin comes in. When he speaks emphatically it is often to combat those who emphatically state that there is no way the sea can possibly rise, with no evidence to support their argument.

"There are real uncertainties and valid criticisms of global climate change, but these guys are way off base," Orrin says. "What we are experiencing, along with the rising sea, is a tsunami of anti-intellectualism. Science is at a new low in the public's view. Scientists are not respected as we once were, and we are out of our league when we compete with the sharpies, the good talkers and salesmen types. We'd rather be out in our labs or out on our research vessels. I think the coal and oil companies, aided by politicians, have done fundamental damage to science in this country. It's true we are not always right. But we deserve to be listened to."

It is no coincidence that legislators who deny sea level rise have made it their mission to dismantle the fine public education system that North Carolina spent many decades creating. Because, on a deeper level, it is education, knowledge, and science that are their true enemies. The developers and legislators are against observation, thought, deduction. They deny not just the fact that the sea is rising, but Enlightenment values themselves. They tell us not to believe what we see. They tell us not to trust our scientists, or to trust only the fringe scientists who happily support what is profitable.

What chance do these higher values have when pitted against personal profit? The developers, and the legislators they have bought, stand to lose two thousand square miles of developable land if they admit the sea will rise. So like the mayor in *Jaws* as the Fourth of July crowds swell, they tell everyone that the water is fine, that there's no shark, and that we should keep building on the beach. And we keep building, and denying, swearing the sun revolves around the earth.

—

Meanwhile the wheel has spun and the arrow has pointed.

We in the Carolinas can breathe a sigh of relief for now.

Those in Louisiana are not so lucky.

THE BIRDS OF
BRITISH PETROLEUM

We are flying into the land of blue roofs. In this place of renewed wreckage, the view from above is vivid and startling. Down below the roofs, half of them cracked open and covered with bright blue tarps, look like a board game of disaster.

Nina and I have not come to New Orleans to do a postmortem on Hurricane Ida, but to attend the wedding of a former student of ours. We planned the trip months ago and had no idea we would fly into the aftermath of an assault.

And an assault it was. Ida was both an echo of the past and sign of the future. With a Category 5 taking aim at Louisiana on the exact same day that the state's most famous storm hit sixteen years before, the word *Katrina* was on everyone's lips. But what had come roaring up out of the Gulf with 150-mile-an-hour winds was as much harbinger as reminder.

For the city of New Orleans itself, Ida, just six miles an hour short of being only the fifth Category 5 storm to make landfall in the United States, might not have even been *the* Big One. But it was certainly that for the residents of Grand Isle, where the storm made landfall, and for those who lived in the swath of land and towns to the west of the city all the way up to Baton Rouge. Only a slight swerve in the path at the last minute spared the city of New Orleans an even more violent assault.

It was plenty bad, of course. Second only to Katrina as far as overall damage. The ninth named storm, fourth hurricane, and second major hurricane of the 2021 hurricane season. Ida came on fast, originating from a tropical wave in the Caribbean on August 23. Six days later on August 29, the sixteenth anniversary of Hurricane Katrina, Ida made landfall near Port Fourchon, Louisiana, devastating both that industrial port and the beach town of Grand Isle. After ripping through the state of Louisiana, Ida weakened steadily over land, becoming a tropical depression on August 30 as it turned toward the northeast, where it would have a few more lethal surprises in store.

Down below is the city, and I can't help but remember the rapid devolving of civilized behavior during Katrina, the stripping away of the thin veneer once power, shelter, and food grew short. Not just the headline-grabbing stories from the Superdome, but the Danziger Bridge shootings, where cops shot and killed two Black men, wounding four others, while later claiming falsely that they were just returning fire from the men.

There are those who have argued that it doesn't always go that way, and I agree. It's not just a cliché: disaster really can rally people, bring out the best in them, bring communities together. To a certain point. But the strain starts to show quickly. I thought of the looting in my own town after Florence or in Mantoloking on the Jersey Shore after Sandy.

And yet here we are, a month and two days after Ida hit, attending a wedding in a city where disaster is the norm and life goes on. It is the way of the world these days. In the midst of relentless tragedy we carry on with the rituals of so-called normal life. We have become inured to this fire, that storm, yesterday's tornado. Those of us who survived the latest round carry on. Blinders to frequent loss are a necessity of contemporary life, of course, and Nina and I wear them, along with our facemasks, as we meet nearly a dozen former graduate students of ours at the Delachaise Restaurant on St. Charles Street in Uptown New Orleans.

Despite everything, I feel real joy in being back in this city. When I first came to New Orleans, to report during the BP oil spill in 2010, it was to cover a tragic event. But I also felt like I had found my lost tribe of eaters and drinkers. "Excess is preferable to deficiency," said Samuel Johnson, and it is hard to find anyone in this town who would disagree with that sentiment. That disaster proved a template for others I later reported on: learning to love, and make merry with, the people in the wounded places I had come to write about. It's the end of the world as we know it and…well, you know the rest.

And it's true. The food and drink taste just as good, maybe better, in the wake of disaster. A vaccination card is required to enter the Delachaise, but since we choose to sit out on the patio the masks soon come off. These are our former students but also people we love. We are told that our students should not be our friends, but, I'm sorry, they are. Teaching is another thing that has changed since the pandemic, and it isn't just the rise of Zoom. Teachers are wary about getting too close to their students, fearing not just infection but the possibility of saying the wrong thing and being called out, potentially canceled. This, as you can imagine, hinders good teaching. And creativity. And fun. It is hard to be relaxed and at your best when stepping through a minefield. Meanwhile the students have begun to act more like consumers, as they make clear with their demands and on their increasingly picky evaluations, and rather than adapting to various styles of teaching from various teachers, they insist on their one way. This, too, is a product of our times, of the dogmatic left reacting to a dogmatic right, and of a way of communicating where humor and forgiveness of foibles seem passé.

This particular group, graduates from a few years ago, may be the last with whom drinking and joking and simple fun are possible.

I push it, of course. The New Propriety makes me uneasy. When I raise my glass for a toast, I ask them a question: "Do you know that I once had sex with a grad student in our program?"

A former student to my left gasps and a couple more look

startled but the rest just smile, in on the joke. They know that after Nina and I moved to North Carolina with Hadley in 2003 my wife decided to go back and get her degree in the program where I was teaching.

I gesture with my glass toward Nina and they all get it, while I get the laughs I was angling for.

Nina is the author of eight books and is now a professor in our department. Many of the students assembled were in her full-year novel class, and one of those students is the groom this weekend, Adam Gnuse, whose final project for that class became a fine published novel called *Girl in the Walls*. If teaching has its downsides, this is the upside. The next day at the reception we feel like proud parents.

It's funny how you don't have to go out of your way to encounter disaster these days. The original plan was for us to fly back the following day, Sunday, but after the hurricane hit I tacked on three more days to my trip so I could continue my explorations of the end-times. After I drop Nina at the airport, I change hats, becoming, once again, DISASTER MAN.

If it is the apocalypse I am after I have come to the right place. Parts of the trip on my drive down to Grand Isle would not be out of place in the pages of *The Road*. It's a landscape in tatters. At Destrehan I cross the Mississippi and head south, essentially following the hurricane's path in reverse into Lafourche Parish, which took the storm's most direct hit. Only weeks ago I was in Paradise, California, and I don't need any reminders about what the future might look like. But if I did, here they are. Enormous live oaks, now dead, have been torn from the ground, uprooted, and thrown across the land like pick-up sticks. Their muscular branches have cracked open roofs and taken out fences. Like dead warriors they sprawl across front lawns. Interspersed with natural debris, the ugly confetti of garbage strewn everywhere, the residents too tired and overwhelmed to begin to clean. Or maybe they are just prioritizing. Blue tarps, the new regional flag, cover not just roofs but a fire engine parked in front of the station and a car wash and a hotel, and boats have been tossed up on the side of the road, including the *Kristen Grace*, a tugboat that now rests on a hill beside the highway. Over in the bayou I see roseate spoonbills and egrets and the roof of a red Honda, the only part of the car not submerged. Signs of debris. The great garbage of the world. And yet this is all just a warm-up for what I will find further south.

The road thins and I cross miles of bridges that arc above water and watery slivers of land before descending into a low, wet world. I get as close as I can to Port Fourchon, the southernmost port in Louisiana, the point where the storm made landfall. Port Fourchon has a strong link to the area's last famous disaster, over ten years ago and five years after Katrina. If you are looking for a connection between our fossil fuel use and storms and oil spills, this is the place. Less a town than a resource colony, Port Fourchon reminds me of western fracking towns like Vernal, Utah, and is not too different from those you could imagine colonizers building on Mars. This particular colony serves as the handmaiden to oil exploration in the Gulf of Mexico, servicing over 90 percent of the Gulf's deepwater oil production with six hundred platforms within a forty-mile radius. Which means this futuristic place that looks more like a factory than a city is the capital of a province that provides close to 20 percent of all oil production in the United States. They won't let me in, but it isn't hard to imagine the mess that the hurricane made of the place, and the uneasy mix of oil and water that flooded through this marshy land.

"Essential Personnel Only" reads the blinking sign at the entrance to Grand Isle, the beach town that was all but wiped out by Ida. I pull over just short of town, at Elmer's Island, to walk out on the marsh, which seems to have fared pretty well. I sketch birds in my journal, including a less-than-shy green heron. When I get back to my car, I ask the construction workers repairing the bridge about the essential personnel sign and they tell me to ignore it—"Everybody does"—so I take their advice and drive right onto Grand Isle, waving to the parked cop as I pass. I feel *essential* and maybe I am. Someone has to bear witness.

If Port Fourchon is a resource colony, this place is, or was, a sight more familiar to those who know the coasts of the American South: a beach town up on stilts. Grand Isle, a barrier island with about 1,500 residents, is technically in Jefferson Parish, which begins in New Orleans before hopscotching down through the islands of Barataria Bay to land here. The homes on this island are the sort that give Orrin Pilkey nightmares, and they have been repeatedly flattened and rebuilt throughout its history, with some of the highlights being Hurricanes Flossy, Betsy, Lili, Isadora, and of course Katrina. As bad and as often as it has been hit, it has never been hit harder than this. The last time I was here there was a sign after the bridge that said, "Jesus Christ Reigns Over Grand Isle." This time around neither Jesus nor his sign are anywhere to be seen.

What I see instead is the work of the hand, or the arm, of the angry drunk. These are not mildly injured houses where mold remediation is the pressing issue. Though a few blue tarps flap in the wind, we are beyond tarps here. These are the shattered and splintered remains of 150-mile-per-hour winds. Houses that have been torn up, ripped apart, decapitated, flattened. Some of the houses slump on the sand, done and devoid of hope of recovery, while others reveal their innards—a fridge, a bed, a couch—while others still remain proudly up on their stilts but wobbling unevenly, like boxers about to fall. The Hurricane Hotel looks battered, as does the Realty Office, which is still advertising deals on nearby homes. Sand covers the only road through town, Route One, and as rain starts to come down that sand starts to take on the texture and feel of mud. I remember visiting Monkey River Town in Belize after Hurricane Iris, where all the formerly paved streets were sand, and this feels just like that. Primitive times have returned. As Billy Bragg sang, the Third World is just around the corner. A few residents pick idly through what remains of their homes.

It is a shattered world, and as such it is itself but also a vision of other places, future places. As more people crowd the earth, as fewer places are inhabitable…well, you get it by now. This is what we have dodged back home. So far. I try to imagine Wrightsville Beach and Wilmington like this, after the angry drunk has ripped through. It isn't hard.

The For Rent signs are a dark joke. To say that it looks like a bomb went off is cliché but no exaggeration. Several bombs. The million-dollar question is how soon until the rebuilding will begin. To rebuild here is absolutely insane—it should be turned into a park, a beach buffer, anyone can see that—but who will doubt it will be done, and fast. Just like the Atlantic beaches, just like the western foothills. Money is at stake. Ah, money! The eager beavers will soon be at it. People like living where they can stare out at the water, even if that water is full of oil platforms and even if it is the kind of water that periodically rises up and flows through your home. For the time being, however, this is the province of out-of-towners. Men from other parishes and from Texas and Wyoming in big white trucks have arrived to clean up the mess before they move on to the next mess. The same folks from Paradise. One of the few economic sectors that is booming these days. The disaster business.

I pull behind some trucks and walk out to the beach. The Gulf is gray and roiling, dotted with the oil rigs that Port Fourchon services. I watch a plover run atop a sandbag. A willet takes off with its seesawing cry. Cormorants fly by. I leave Grand Isle without talking to anyone, which is not like me. But there is something too creepy about trying to interview the people picking through the remains of their homes. Too intrusive. Also, except for the few home pickers, no one here is *from here*. I get in my car, pull a U-turn in the mud/sand, and drive off the island.

—

It is all tied together. We know that. The oil. The marshes. The storms.

Spill. It's such an innocuous word. As in, "Oops, I spilled."

The first time I visited Louisiana was in April of 2010, the year the BP Deepwater Horizon blew, killing eleven men. For the next eighty-seven days hundreds of millions of gallons of crude oil flowed into the Gulf. By late June, back in Wilmington, I had grown sick of other people telling the story, so I picked up an assignment from an environmental magazine and headed off to see the disaster for myself. I was told that the beaches were all closed, that no one would talk to me, but I found the opposite. People were dying to tell their stories. Pun intended. For the next month I toured the tar-balled coasts, hitched a helicopter ride with the Cousteau film team to see the rig and the giant oil slicks, and listened to local fishermen, scientists, rescue workers, and politicians while observing the media circus that I was part of. Traveling along the marshes and barrier islands, I had the deep sense that no one knew what the hell was going on: not BP, not the government, not the scientists, not the media, not the stunned locals, and certainly not those who captained the cleanup boats, the Orwellian-named "Vessels of Opportunity."

After weeks in the Gulf, I was feeling somewhat hardened to the catastrophe, but something cracked when I visited the Fort Jackson Oiled Wildlife Rehabilitation Center, a kind of avian MASH unit on the front lines. There I stared into the eyes of the very symbol, even the cliché, of the spill. The oiled pelican! The mascot of that grim disaster. The brown pelican that gazed back at me from inside its cell of a plywood box had wings still dark with oil. Face-to-face with this bird, I was struck by the deal we had made as a culture and a society, surrendering animals and destroying their habitat, all so we could keep living exactly as we do. I didn't in any way exclude myself from the accused group. I stood and stand accused.

But seeing into the eyes of that bird I was reminded: We are not just doing this to ourselves. We are doing it to them. To the animals. To their homes. To everyone and everything on earth.

Five years later I returned to the Gulf to see how the birds were doing. I had come back to write a sort of avian state of the union report for *Audubon* magazine, and one of my first stops was the southernmost point in Louisiana. If I am honest, I came to my work full of prejudice. I was not a good, or at least impartial, reporter. I was there to write a grave environmental report on the death and doom of all things avian. My plan was to report on how bad it was in hopes of making it better.

The trouble was that no one had told the birds. Rather than paucity, I found at the southern tip of the state in the aptly-named town of Venice, on a causeway of a road that led straight into the Gulf, a scene of wild abundance as thousands of birds from dozens of species appeared to be swept up in an ecstatic celebration of sunrise. Though they didn't know it, they weren't doing themselves any favors. Within a hundred feet of the road anhingas and cormorants and loons were diving and an osprey was scaring some turkey vultures out of a dead cypress while dozens of roosting brown pelicans, a tad slower to rise, filled the dark silhouetted trees. The rising sun turned the mob of white ibises pinkish and a roseate spoonbill even pinker, while lighting up the undersides of the white pelicans that wheeled above.

My friend Mark Honerkamp came with me during that trip in 2015. Mark, who had been part of so many past adventures that he was threatening to have a card made that read "Professional Literary Sidekick," could barely stop laughing at the symphonic craziness of so many birds flying in so many directions as we turned our binoculars one way and then another. Look, dozens of black skimmers carving the sky over Barataria Bay. And Jesus, over there's a marsh hawk and a merlin; and over there, in that puddle-like pond, a king rail, an avocet...

I don't want to give the impression that we were in the heart of some pristine wilderness. We were on a road after all, a causeway that stretched across the wetlands, where we could see the eternal flame of a gas flare above an oil refinery, a hundred-foot column of fire licking the sky, which provided a kind of twin to the fiery ball of the rising sun. When you got too close to the roaring refinery you couldn't hear yourself think let alone a heron sproak. Furthermore, the place where we had spotted our first ibises was off of Haliburte Road, an indicator that birds weren't the only species that found what they needed in these waters.

Soon after sunrise we met David Muth, then the Louisiana State director of the National Wildlife Program, and were joined by a quiet but efficient birder named Lisa Stansky, plus James Beck, who at first glance looked more like someone who would be here to *shoot* the birds but turned out to be a kind of brilliant savant gifted with supernatural eyesight and hearing. We were joining them for a Christmas bird count and we were sensible enough to just listen and not say a word. Before we could lift our binoculars, they pointed out this black-necked stilt or that neotropic cormorant or those black-bellied whistling ducks, and by morning's end we had added several birds to our life lists. Our tour ended at the Venice dump, where gulls swirled in a great cyclone above a tractor pushing a mountain of trash and David, letting me look through his telescope, showed me a rarity, a single glaucous gull sitting among the thousands of other white birds that, to my eye, looked no different.

"Of course even by conservative estimates all this will be underwater in less than a hundred years," David said.

He extended his arm like a game show host and I understood that "this" meant not just the bird habitat and the harbor but the oil refineries and the dump. He wasn't exaggerating. The official elevation for Venice is three feet *below* sea level, and you can stare up toward the levee at enormous ships on the Mississippi that appear to be floating above the town. There was little left of this place after Katrina, and though it dodged a bullet with

Ida one suspects its rebirth will be brief, though few say this out loud. This is bad news for the humans, of course, and also for the birds. The Gulf wetlands have long been disappearing due to subsidence, the land sinking as the seas rise at a rate as fast as any in the world, and all the oil has exacerbated and sped up the process. The birds need places like this.

During that same trip Mark and I also visited Grand Isle for the first time. Like Venice, it is one of the state's southernmost points, which means it is a jumping-off point for many migrating birds in the fall when they head down to Central America and a resting point when they return in the spring. On a December day that seemed more arctic than tropical, Erik Johnson, then the director of conservation for Audubon Louisiana, gave us a tour of the beach at Grand Isle. Cold winds strafed the beach as we stared through our binoculars at various plovers—semipalmated, piping, and snowy—while Erik filled us in on the state of the Gulf birds.

"The wrack line was full of oil after the spill and so the volunteers had to rake it away," he said. "But of course the birds need the wrack line. It provides insects, and a windbreak."

He explained that in 2012 Hurricane Isaac had created a kind of rerun of the BP spill, oil churning up all over again. Which made sense. While there is much we still don't know about the spill, one study has confirmed that six to ten million gallons of oil remain buried on the ocean floor. The mystery of where the oil went is really no mystery at all: it was, with the help of chemical dispersants like Corexit, swept under the ocean's rug.

Out on the water, beyond the line of birds along the shore, a dozen or so oil rigs stood on their stilted legs, looking like abandoned space stations. Erik spotted a herring gull whose face was dirtied with oil. He called the oil hotline on his cellphone and was transferred to several different offices. He told us that reporting oil is always tricky business, a byzantine affair where you are never sure if your call is ever really recorded.

As we stood on the beach the rigs shimmered on the horizon. The oiled gull was a reminder that not all of the oil out in the Gulf was BP's, and that a small spill's worth of oil is dumped into the Gulf almost annually. Little had been done to slow or regulate the drilling, the 2010 moratorium on drilling ending well before that on fishing and shrimp. Another thing that most people knew but didn't say out loud was this: another huge spill was surely coming.

Erik was one of the people in charge of compiling the Louisiana Christmas bird counts. I mentioned the study that came out the spring before, coauthored by Chris Haney, chief scientist for Defenders of Wildlife, who previously studied how the Exxon Valdez spill affected birds and whose paper attempted to offer an estimate of the BP spill's cost to regional birdlife. Haney concluded that approximately eight hundred thousand birds had been killed by the oil, despite the fact that fewer than eight thousand cadavers were collected and counted in the year after the spill.

"Of all the spills I've ever studied," Haney had said to me, "none has ever had as many combinations of factors that have made it harder for a dead bird to actually reach the morgue and be counted."

Those factors included the fact that only the BP-approved could pick up the birds (I was once in a boat with a charter fisherman who said he was sorry but it was against the law for him to pick up an obviously oiled and distressed heron), that the oil had spilled so far off-shore and then spread over 2,500 square miles, and that gulf currents could whisk away the dead bodies, the perfect crime. Still, I wondered about one part of the study, which said there had been a major reduction in the population of Gulf laughing gulls, basing that conclusion on Christmas bird counts.

Erik nodded when I told him this. Then he hesitated, as if not sure how much to say.

"So far the Christmas bird counts don't really reveal a reduction in birds. I'm worried that the so-called 40-percent reduction in laughing gulls was the result of a cold, windy day, kind of like today, when the counters saw very few birds overall."

Erik was in no way saying that, to paraphrase The Who, the birds were alright. Just that the state of the science was, kind of like the Gulf itself, a mess. He was one of the people trying to change that by compiling the Christmas bird count data, which, as Chris Haney and others pointed out, *can* reveal trends over the long haul. Erik reiterated the "no baseline" theme I'd been hearing throughout my trip. One of the challenges with figuring out how the oil spill, the subsequent hurricanes, and the massive loss of wetlands are affecting the birds is the lack of a database. Even before the BP disaster, the Gulf was a region of neglect. We certainly have not treated it like a spot that deserves to be studied, which would have been helpful. Many scientists say it's practically impossible to determine what the state of the Gulf ecosystem is now because we didn't know what it was *then*.

For offshore birds in particular, the amount of monitoring data is "mind-numbingly sparse," according to Haney. "The Pacific Coast is well-studied, while the Atlantic, by contrast, is kind of an orphan. But if the Atlantic is an orphan, then the Gulf is like a baby abandoned on the church steps."

We headed to Elmer's Island, where Mark caught four redfish and saw more birds than he would all winter back in Boston. Once again the Gulf's abundance was quick to undermine, or at least counterbalance, thoughts of doom. Erik had a long drive back to Baton Rouge and said goodbye, but we decided to stay and keep fishing. We watched as the dying sun shot streaks of orange down into the water. The trees darkened to silhouettes.

Mark cast his line into what now looked like a creek of molten lava. Mergansers and buffleheads floated further down the creek, and we heard a clapper rail breaking out into what sounded very much like applause for the setting sun. Out in the

Gulf the lights on the rigs came on, the structures sparkling like Christmas trees. I knew that all the oil still out in the ecosystem continued to have effects, both lethal and sub-, both known and unanticipated, and that, combined with ever-shrinking habitats, did not bode well for the birds of British Petroleum.

Still, the Gulf remained a place of wild abundance, a place where Mark no sooner dipped in his line than he caught another redfish, and where we, even as the sun finally dipped below the horizon, were surrounded by a dazzling variety of birds. Despite its strangeness and its many contradictions, I could feel myself once again falling hard for that doomed and dying place, that living embodiment of the clash between our need to consume and our love of beauty, the very place where our possible futures are fighting it out.

—

After the devastation of Grand Isle, I make my way north. I'm looking for hotels but they are either storm-damaged and closed or filled with people left homeless by the storm. It isn't until I am past New Orleans that I find a room.

The next morning I continue south. Before I came here in 2010, in my early ignorance, I had imagined New Orleans *on* the Gulf, not understanding that you can drive below it on a south-eastern slant for another seventy or eighty miles. My first stop today is in Lower Coast Algiers, a good ways down the Mississippi but still technically a part of New Orleans. I've come to case a joint: the house where Adam, the groom of the weekend's wedding, grew up. I am not here to see him or his parents, but to do a little literary sleuthing. I'm reading his novel at the moment and it all takes place inside of a fictional version of the house I am now staring at. The house is elegant, almost dainty, with many gables and blue trim, though not as big as its fictional counterpart. The book climaxes with a hurricane, and it isn't hard to see why. From the edge of the front lawn you could throw a baseball up onto the levee. Adam's family evacuated during Katrina and returned to find that barges had broken free during the storm and one had lodged itself atop the hill above their house. They were lucky though: if the river had been a couple feet higher it would have rushed through their living room.

After I take a couple pictures of the house, I find a spot to park across the street and climb the green hump of hill that protects Adam's family and the rest of the parish from the mighty Mississippi. *Levee* is a common enough word, but in my experience many people from other parts of the country have only heard the word in songs and often don't really know what a levee is. What it is is an artificial hill, usually covered with grass, built along the river to keep it from flooding. In theory. Often enough during storms levees are topped.

At the edge of the water I take a seat in the shade of a cypress. The strip of land where the river laps the shore on the other side of the levee is called the batture. According to the law library of Louisiana, the word *batture* comes from the French word "to beat," referring to the alluvial land that is "beaten" by the river. From here I can stare upriver at the city or across at the inevitable oil refineries. A barge moves sluggishly down the river. A white ibis flies across its bow.

Things grow less quaint as I head south. I enter Plaquemines Parish, where I stayed during the summer of the BP spill. Again I am reminded of Vernal, Utah, where the main business, like here, is the getting of energy. Though I am also heading toward some of the most famous fishing grounds in the United States, there is an underlying factory toughness to the place. Roseate spoonbills and egrets blend pink and white against a stormy sky as I pass Alliance Energy, with its giant vats and refinery fires licking the underside of the clouds. A mile further you can see how Ida has thrown the marshes of Barataria Bay up over the road. Work crews with orange vests clear the clumped stalks and grasses.

It is right here, at the sixty-one-mile mark on Route 23, where dozens of ibis pick among the strewn garbage, that you can see both the most pressing problems for this region and a potential, if partial, answer. The land is ridiculously narrow: on one side of the road is the levee with the Mississippi lapping behind it, on the other the waters of Barataria Bay. Not long ago Barataria was a famous fishing ground with a rich ecosystem of deep water and marshes and islands of dry land, but over the years, through storms and subsidence and being carved up with straight-lined oil company canals, it has all but disappeared. Land from when I was last here is now water. This is one of the places where the famous football field metaphor often used to describe land loss in these parts is most apt, as in "we lose a football field of coast every hour." During my last visit they were predicting that the marshes of Barataria Bay would be wiped out. Now there is no reason to use the future tense. Now they are lost. Ida was the death blow.

Another mile and I come upon Ironton, a Black community that was devastated by Ida. For the most part Plaquemines Parish got off easier than Lafourche, but Ironton was one of the exceptions. I pull off the main road and drive down the causeway toward the little town. More a settlement than a town, really. The entire place is made up of four streets and a hundred people and almost everyone has lost their home.

Even as I drive in on the causeway I can see that houses have been flattened and shattered. I will not be getting any closer, however, because a sheriff gets out of his car and blocks the road. He is a man of few words.

"You can't come in here," he says.

No gawkers allowed.

"It was bad?" I ask.

"*Bad* bad."

Ten minutes later I pull over and call Faye Matthews. Fayenisha Matthews is senior counsel at the National Wildlife Federation, where she works on legal and policy coastal land protection and restoration. From Faye I learn just how bad *bad* is. No electricity, most of the homes completely destroyed in a place where few have the means, or insurance, to rebuild. No water either, except for the kind flooding their houses. The residents feel lost and abandoned.

I have some sense of this already thanks to an interview I listened to with Pearl Sylve, an Ironton resident, who told WDSU, a New Orleans station: "I have heard all the places. Houma, Lafitte, Thibodaux. I have heard about all the places except for Ironton. We're a small Black community. Everyone back here been here for years. No one thinks about us."

Sylve evacuated before the storm and came back to find her home was no longer where she had left it. It had crossed the street and was now in her neighbor's front yard.

"My house floated from the Johnson side of the street to Glenda Green's house. That's Glenda's house," Sylve said, pointing.

A ten- to fifteen-foot storm surge had breached the levee. And ten to fifteen feet of water then flowed through the homes of Ironton.

"Go and come back," Sylve said. "That's what we always do. Usually you have something to come back to. This time we really have nothing to come back to."

—

My next stop is the small town of Buras and the lodge of Ryan Lambert, who was my main source when I was reporting on the oil spill and who became the main character in my book about the disaster. Ryan and I are at the opposite ends of the political spectrum, and during the spill he would tease me and refer to Obama as *your* president. Back then Ryan's vast lodge was almost empty since he refused to rent out rooms to the men who had come from out of town to man BP's so-called Vessels of Opportunity. By the next time I visited, for my bird report for Audubon in 2015, the place was jammed with hunters and their dogs, and you could practically smell the testosterone. It was a room full of alpha males but there was little question who the alpha alpha was. It has been six years since we have seen each other and I can't say I am unhappy we skipped the Trump years entirely. But while we don't agree on much, our common ground has always been the ground. Or, as the ground is now known locally, the water.

When I saw the first projections for Ida, I thought *Oh no*. My mind raced to Ryan and the lodge. His lodge had been the absolute bull's-eye for Katrina and the watermarks from that storm were up to the rafters. Which means that during the hurricane his lodge was under twenty feet of water.

The day Katrina hit had been a long, tragic, and heroic one for Ryan. He wasn't worried the day before, and had been out fishing, but then Katrina veered and suddenly started heading right toward him, as if, he said later, it were seeking him out. With no choice, he evacuated and headed north to the city, rescuing an elderly Vietnamese couple, whose truck had broken down, along the way. Then his sister called and told him their uncle Rich had a broken ankle and was trapped in the city with the waters rising around him.

All the cell phones were out because the towers had been knocked down, but oddly some landlines were still working.

"I'm coming to get you," Ryan said to his uncle. "I have no idea when I'm going to be there. It might be midnight or it might be two in the morning. But just know I'm coming. Just go upstairs because if you get in that water you're done."

His wife told Ryan he couldn't go rushing into a flooded city where there was already panic and rioting and looting, and where power lines were falling into the water. He said, "What do you mean? This is what I do for a living. I hunt and I guide. Now I got a license to do it." He grabbed four or five guns and hitched up his boat and for good measure threw his bicycle in the back of the truck.

Once he got to the city, Ryan talked his way past a cop onto a closed highway and then made his way into the drowned town atop the levee, dipping back down onto the flooded road whenever he hit an obstacle. He cut up Causeway Boulevard and started heading toward his uncle's house in Metairie. He couldn't get any farther in the truck and so he grabbed a couple of guns and waded on. When he got to the house, he yelled up and banged on the door. "Bogale, that you?" his uncle asked. The old man's truck was parked on high ground and he threw the keys down for Ryan. Ryan backed it up to the window, where his uncle could climb down along a pipe into the truck bed. They drove slowly out through the submerged and deserted streets, the truck kicking up a substantial wake, and ditched the old truck for Ryan's once they reached it. Then it was back up on the levee. They made it home about four in the morning.

Weeks later, when Ryan and his neighbors returned to Buras, they found rubble.

"People would come back with trucks and trailers to look for their stuff," Ryan told me. "And they would leave with a baggie."

This fall, as Ida churned across the Gulf, it was starting to look like a rerun.

As late as August 27, two days before landfall, it seemed Ida was going to follow the exact course of Katrina. Then came the slight swerve that saved him. So much depends on small changes. Good for Ryan, bad for Grand Isle.

Watching back in Carolina, I felt relief.

But what I don't know as I pull into the driveway of the lodge is that this time around it was Ryan's home up in New Orleans that was hit. For six hours he had been in the eyewall, the water on a nearby lake flying sideways and the trees looking like they were being shaken by giant hands. Still, his house had done better than the others in the neighborhood, which meant he now had fifteen people, all homeless neighbors, living with him.

Today the lodge is empty except for Ryan. The two of us sit

at one of the dining room tables in the cavernous building. The lights are off, the late-day sun slanting in through the windows. Ryan looks tired. He hunches over the table.

Before we start to talk the phone rings and he gets it.

"I am bulletproof," he says to the friend who has called to check in, but I am not so sure. He is a tough guy but one can only take so much. "These fucking hurricanes," he sighs into the phone.

When we met I was forty-nine and he was fifty-three.

Eleven years have passed. Neither of us are spring chickens.

"It sucks getting old," I say when he gets off the phone.

"Tellin' me."

This morning, as a kind of therapy, he headed out for a day of fishing alone on Barataria Bay. This was unusual for a man who guides others for a living. It was also unusual that in four hours he, considered a top fisherman in one of the world's most famous fishing grounds, caught only five redfish. If he was trying to get away from the aftereffects of disaster he had gone to the wrong place.

"There's nothing left out there," he says. "The whole estuary's caving in. It used to be eighty, ninety feet deep. Now it's ten feet deep. Because the marsh, as it dies, sinks to the deepest spot and fills it in. All the material sinking to the bottom as it decays takes up all the oxygen. The deep holes are gone. What used to be the best wintertime fishing grounds in North America is now a dead zone."

Understandably, it was human losses that made headlines after Ida. But the story that did not make it into the news was the death of Barataria Bay.

He shakes his head.

"We've already lost 2,400 square miles. That's bigger than Delaware."

Ryan has been hunting and fishing these waters for close to fifty-five years, but maybe even he doesn't understand how vital he, and others like him, are to our understanding of what is going on in the Gulf. He knew this place *before* Katrina, *before* massive erosion, *before* the spill. And if there is one thing missing in our attempts to understand the loss in the Gulf, it is the lack of *before*. The lack of a baseline. This is a point that every scientist I have talked to has hammered home. Our baselines have shifted to the extent that we forget the bounty this place once held, and so accept something less. Much less.

In 2015, when I came back, I asked Ryan about the duck population, since that was his worry right after the 2010 spill.

"The ducks are back," he said, pointing around the lodge. "That's why this lodge is full. We've had a couple of our best duck seasons over the last couple of years. The trouble is they are not breeding here. In 2012, Hurricane Isaac dredged up more of the oil and we lost more wetlands."

Habitat loss. That is the real worry if you care about the birds here, whether your interest is in watching them or shooting them. There are times you can't be sure of much in the Gulf, but amid all the confusion, one thing is undeniable: habitat is going away, and it is going away fast, the land sinking and sea rising like almost nowhere else on earth. We are not talking about geological time here but about whole marshes and small islands that have disappeared since I last visited.

We have our political differences, but I admire Ryan Lambert. His is a buoyant nature. If resilience is a watchword in all these communities leveled by disaster, Ryan is resilience personified. He embodies a kind of toughness and practical know-how that is better than mere hope, and that, along with some good luck with his bad, has allowed him to survive so far.

Within ten minutes he has moved beyond despair. What's the point? Rolling up your sleeves works better. He pulls out his laptop and goes into teacher mode. He takes some glee in the fact that, while I am the professor, he is the one usually doing the professing when we're together. Today my teacher has an important lesson for me. He wants me to understand that, despite the oil, despite Katrina and Ida and all the other fucking storms, despite the subsidence, despite everything, all is not lost in this part of the world.

"The patient's right there," he says, pointing west toward Barataria Bay.

Then he jerks his thumb behind him, to the east: "And the doctor's right there."

The doctor in this case is the Mississippi River. And what Ryan wants to do is simple. Connect the doctor with the patient. How will he and others do this? By creating redistribution channels that funnel parts of the river, and more importantly its sediment, into the bay. This will in turn build up land in the places where it is disappearing.

In other words, in this often depressing place, there is a chance. During my travels I have been wary of the word hope. But this is different. This is not Disney hope but something as real and solid as the land that Ryan has already helped create.

It is back at mile sixty-one, the part of the highway where the marsh washed over, a mile north of Ironton, that one of those sprigs of hope can be found.

It is there that 1.2 billion dollars is being spent to connect the doctor with the patient. The goal of the Mid-Barataria Sediment Diversion project is to channel water from the Mississippi across to Barataria Bay so that the sediment from the river will regrow the marshes, providing habitat for birds, fish, and other animals while providing a buffer to protect settlements of human beings, like Ironton.

This is no pie-in-the-sky project. The day before I arrived at Ryan's I had reconnected with David Muth, now the director of Gulf restoration for the National Wildlife Association, and he confirmed everything Ryan was telling me.

"This place was a natural wonderland," he told me. "Bison used to roam in the low-tide marsh. The Mississippi River flowed into a series of deltas about six or seven thousand years ago. The first European settlers built the levees. There used to be multiple tributaries but when they were closed off all that sediment was just dumped in the Gulf.

"Meanwhile the seas are rising here while the land is sinking. We are subsiding. It is thirty thousand feet down to the bedrock. As far down to the bedrock as Mount Everest. Instead of bedrock you have clay and silt and sand, and so the land sinks. At this point, no matter what we do, we are going to lose significant parts of the coast. You can't hold on to it. The resources are not there to hang on to the remaining four thousand square miles of wetlands in Louisiana. We are going to lose large sections of the existing delta and coasts. That's a fact. In the face of climate change we are going to lose major parts of the delta. But we can offset the loss by driving productivity in the remaining delta. We can concentrate on a smaller, more sustainable, delta.

"The way to do that is building sediment. If we divert the river it will build up land. We can turn the energy of the river back into the overall system. Right now the sediment is going to the wrong places. Connect it to the delta cycle. Build marshes. Change the trajectory based on the best science."

Given that it has been a decade and a half of historic hurricanes and oil spills, and the fact that a near–Category 5 just blew through town, I was surprised by the optimism in Muth's voice.

But Erik Johnson, who gave me the tour of Grand Isle seven years ago and is now the director of bird conservation for Audubon Louisiana, sounded equally hopeful when I called him.

"Money helps," he said.

He pointed to another irony of this particular environmental fight. None of it could have happened without BP and the many billions that were part of the oil giant's settlements. Five billion dollars in the right hands can go a long way toward a healthier Gulf.

"Broadly speaking, since you were last here the Gulf has

moved from response to recovery. People are on board, most people, and we have a once-in-a-lifetime chance. The BP settlement is an opportunity. And right now it feels like Louisiana gets it. Maybe people get along better when there's lots of money in place. With BP writing the check we can reconnect the river to the delta. The river, and its sediment, could bring the bay back to life."

Hurricane Ida does not render this a secondary concern. It means sediment diversion is needed more than ever.

"Every mile of land you have between you and the hurricane knocks the storm surge down one foot," Ryan says.

Of course not everyone is on board. The oystermen and shrimpers oppose the redistribution because the intrusion of the river will alter the salinity of the water in the places where they have their claims. This does not mean that oysters and shrimp will disappear, but it does mean they might not flourish in spots that particular oystermen call their own. A decent amount of the money spent on river diversion is going into buying out those claims, and still many oystermen, fearful of change, stand in the way of the project. Ryan has little tolerance for them.

"They hate me," he says. "I don't mind. I call bullshit on them. I'm the only one brave enough to do it."

He shakes his head. "Look. It needs to happen. I take people out in my boat and point to where the land used to be."

He shows me maps where land was when I was last here and water is now. I remember the day six years ago when Ryan took me out on his boat and pointed down at the GPS, which claimed, despite the evidence of open water, that we were floating on top of six miles of grass. There wasn't a single blade in sight. And this disappearing act has only accelerated over the last six years.

But Ryan believes this can be reversed, and he has the proof. He is no theorist, and for him river redistribution is not theory. In fact, he has already taken matters into his own hands. While the slow-moving wheels of government might be spending over a billion dollars to create the Mid-Barataria Diversion, he has already created a few all on his own.

Most recently he, working with Ducks Unlimited and several government grants, created a small diversion on the east side of the river that feeds a four-thousand-acre area that was water before. Ryan's newest business is land-building and the results are miraculous. Some of the land-building has happened in a matter of months, with such speed that people didn't believe him. Not long ago he had brought a member of an environmental organization out in his boat to inspect the new marsh and said, "Jump out here." She wouldn't, assuming she would sink to her waist. But when he finally convinced her, she found herself standing on solid ground.

"Nature is so resilient," Ryan says. "If we just let it work for us."

Even on this smaller project there were plenty of obstacles. If the need to build land seems self-evident, what is self-evident, per usual, runs right into self-interest. In this case it isn't oil executives but the oystermen again.

"They hate my guts. The feeling is mutual. It cost me $75,000 to buy them out of their oyster leases even though they hadn't had an oyster there in twenty years. It doesn't matter to them if there are any fish or land as long as it is good for the oysters."

Another enemy, one familiar to me from my time with Orrin Pilkey, is the Corps of Engineers.

"The river has been running high for nine years," Ryan says. "The Mississippi is the most engineered body of water in the world. And all the locks and dams on it are now filled with sediment. That's because there are no doors to release it. So all the sediment that used to come down here is stuck behind those locks and dams.

"The Corps of Engineers has got this country so screwed up. If they don't have a reason to have the locks and dams for navigation anymore, because there's no more lumber industry because we've used up all the lumber, then get rid of them. If they're not using them for creating hydroelectricity, get rid of them."

Meanwhile the sediment that does get through the dams is mostly dumped at the mouth of the Mississippi, where it falls off the continental shelf.

Ryan shakes his head. *What a waste.* All that good land-building material gone.

As for Ida, he says: "We got ungodly lucky down here. I feel guilty about it."

Before I leave, he shows me a video on his phone of the wild shaking of trees by his house up in the city during Ida.

"What are you going to do about the thirteen people now living in your house?" I ask.

"They're going to get the hell out."

A long pause, then, softly: "Next year."

—

I have sworn off the word *hope* during my travels. It seems overused, trite, the kind of thing that must always be tacked on to the end of any climate change story in the news.

I drive farther south. It is a startling sight, if you have not taken this road before, to suddenly be driving parallel to an oil tanker on the other side of the levee. It's hard to believe there is more south from here, but there is. The land thins, attenuates. Soon every road feels like a causeway and the water laps close. Whole trees are white with egrets. I roll my windows down and listen to the low, industrial hum of the refineries intermingling with the gurgles and clucks of the hundreds of cormorants that perch in the cypresses near the road. As soon as the sun breaks through they spread their wings to dry.

When I reach the very end of the road, the southernmost point in Louisiana, I find the usual mishmash of industry and nature. Not much has changed since I was last here except for the blue Trump flag hanging from the outermost shack. On my way back across the causeway a truck speeds up to tailgate me. Maybe it's a little scarier down here than last time. But maybe it's always been a little scary.

On the way back north I pull into Fort Jackson, a beautiful half-underground fortress right on the Mississippi that is built into the side of a hill below the gnarled limbs of massive live oaks. The fort was built in 1822-23 to protect the lower Mississippi and later fell to Union forces after the city was burned. In 1961 it was declared a national monument. There is not another human being in sight, but about a hundred white ibises peck at the large lawn beside the fort.

When I came here during the spill this was the command center for the bird rescue, and the lawn was covered in tents. Here is where human beings worked round the clock to clean the oiled birds that were being brought in from the Gulf. And it was here that I looked the oiled pelican in the eye and felt something crack inside me.

What both birds and people need is the same: habitat. That doesn't mean that a hurricane won't hit next year or later this season or that another oil spill might not happen at any moment. And it doesn't mean that we, stubborn foolish beings who put ourselves first, won't find a way to undermine the natural corrective, the route to recovery, that is here for the taking.

It would be foolish to be too optimistic in this wounded place, and I don't want to make the standard nature writer move and end on an uptick. But everyone I have spoken to has stressed that in the Gulf, unlike so many places, a degree of recovery is possible. A clear path forward exists and, so far, the people and government, bolstered by BP money, seem committed to taking it. How odd that in this home to the most dismal of environmental stories, this threatened place, this resource colony where oil spills into the sea and the land itself sinks as the seas rise and birds fly backlit by the eternal flame of Halliburton, and the storms grow stronger and hit more frequently and harder, there is a way forward. It almost inspires me to re-embrace that over-used and outdated word, the one I had sworn off forever.

AFTER THE STORM

Phenology is a word that describes the phenomenon of the natural world throughout the year, from the budding of trees to the flowering of plants to the migrations of birds. Multiple scientific studies have shown that the timing of the year in many places has now been thrown off by climate change. But it isn't just that. Soon any reliable almanac of the year will have to take into account the regularity of disaster. Perhaps we will call it catastrophology.

Let's pause and remember that much of the present-moment thread of this book, from July's flashfloods and fires to Ida's hitting New Orleans, is taking place in little more than two months. Fast forward a little bit and you will find the latest first snowfall on record in the front range of the Rockies and Midwestern tornadoes and western fires in winter. And this is just the United States and a small sliver of the disasters that now make up our climate-age almanacs.

We think in old ways, we can't help ourselves. We need new thinking in a new world but we are not there yet. For instance, it was natural for people to think, on August 30, as Hurricane Ida weakened and headed north, that it had already dealt its harshest blows, and that after leveling Louisiana it had done its work. The country's focus was, understandably, on New Orleans. It's safe to say that no one, at that point, was fretting about another metropolitan area, which, along with New Orleans and Miami, is considered by climatologists one of the three most potentially dangerous and disaster-prone in the United States.

That is to say that, after the winds started to head north and inland, leaving a ravaged Louisiana in their wake, and the storm seemed to dissipate, no one, including the climate modelers and weather forecasters, would have guessed that it would claim many more lives and cause catastrophic damage in New Jersey and New York City.

In fact, Hurricane Ida would kill more people in the northern United States than in the South. As of September 15, 2021, a total of 115 deaths had been attributed to Ida, including ninety-five in the United States and twenty in Venezuela. In the United States, thirty-three deaths were in Louisiana, thirty in New Jersey, eighteen in New York, five in Pennsylvania, three in Mississippi, two in Alabama, two in Maryland, one in Virginia, and one in Connecticut.

In the end Ida would leave behind not just death and pain, but also a fable for the new age of climate. It would teach us again that, for all our predictions, nothing can truly be predicted. TV projections like to display the aptly-named "cone of uncertainty," the dozens of potential routes for each hurricane displayed in colorful lines bouqueting their way up the coast. And Ida would remind us that the cone of uncertainty reigns. It would also, echoing Sandy, remind us that what had been primarily a southern phenomenon was now a northern one as well. That uneasy feeling we in the South have when summer rolls around is one that New Yorkers better start getting used to.

—

During my travels with Orrin, he gave me a history lesson or two about hurricanes and New York. In 1821 a Category 4 hurricane hit New York City directly, raising a storm surge of thirteen feet in an hour, cutting the island in half, and flooding the entire city. In 1938 the famous Long Island Express hit the coast with a storm surge twenty-five to thirty-five feet high. Hurricane

Donna hit New York on September 12, 1960, with ninety-mile-an-hour winds and five inches of rain. Orrin added that the US Landfalling Hurricane Project predicts that there is a 90 percent probability that the New York/Long Island area will be hit with a Category 3 hurricane over the next fifty years.

But while several hurricanes have hit New York, it was Hurricane Sandy that ushered in the new era we find ourselves in, and not just because it introduced the contemporary North to a southern storm. Sandy was a game changer.

I had a personal interest in Sandy since, as it happened, my city, Wilmington, was the original bull's-eye for the storm's landfall. In fact, in the early projections it looked like it was making a beeline for our house. I imagined Jim Cantore standing with his microphone and wind-flapping windbreaker behind my writing shack, with our two dogs at the time, Missy and Mandy, barking at him, as I tried to take last-second notes on the birds that were hunkering down for the storm. But the early predictions are never precise, and, lucky for us, the storm didn't do as it was told. Soon the lines had shifted and Cantore had moved on. The storm, which had seemed to be curving up directly toward my home at the mouth of the Cape Fear River, abruptly bulged eastward, changing from an exclamation point into a kind of crude question mark, arcing outward and gradually back in toward New Jersey.

I stayed out in the writing shack that day as the waters rose. At first they formed a moat around my castle, then they entered the shack itself, covering the floorboards until I could slosh through a foot or two of water. Our yellow labs happily joined me. Overall our damage was minimal, other than the minor flooding that briefly changed our home into the lakefront property my wife had always dreamed of. But Sandy wouldn't go as easy on other places.

Like many tropical depressions, this one started small, boiling up from the ever-warming waters of the Caribbean. Sandy began life as a tropical wave on October 22, 2012, but due to low winds and warm water, became cyclonic within hours and by the end of the day was declared a tropical storm and given its official name. Following a trough, a region of low atmospheric pressure, it slowly built up speed but it wasn't until two days later, October 24, that it was declared a hurricane. Later that day it hit Jamaica, the first hurricane to strike that country in over twenty years, tearing off the roofs of most of the buildings on the eastern shore, sending over a thousand people to shelters, and plunging the island into darkness.

Sandy was just getting started. The winds that Jamaica experienced were comparable to those New Jersey would later feel, but as the storm veered north it rapidly strengthened and by the time it hit eastern Cuba the winds had reached 110 miles per hour. (If you have ever stood outside in hundred-mile-an-hour winds you will remember it.) Waves of twenty-nine feet hit the coast, with the storm surge adding another six feet. The eastern end of the island flooded and left a devastated and altered landscape. The beautiful city of Santiago de Cuba, which I had visited in 2005 to study osprey migration, was already built on ruins from an earlier time. It took a direct hit, with almost all of its trees torn up by the roots and nine people left dead in the rubble of destroyed homes.

The aptly named Windward Passage is the body of water between the eastern end of Cuba and the western point of Haiti. It is only fifty miles wide, and as the eye of the hurricane crossed Cuba its eastern flank sent massive waves toward Haiti's shore, submerging large parts of the island. The water of the torrential rains and the waves mixed with the water of rivers overflowing their banks. Thousands were left homeless, fifty-four people died, and cholera, which was already present, worsened to epidemic proportions.

The Bahamas and Bermuda were luckier, as the winds had slowed by then, though they still suffered massive power outages and millions of dollars of damage. That was the point where the storm seemed to be heading our way until it changed its mind and wandered eastward, back out to sea. By then the winds had slowed and were generally clocked in the sixty-mile-per-hour range, with gusts in the eighties. It is hard to imagine what would have happened had winds the speed that Cuba experienced hit the Jersey Shore or New York City. It wasn't wind speed that would define Sandy in the North, however, but sheer size and duration, as well as the impeccable timing of high tide and full moon and a simultaneous storm to the north. In the United States, Sandy would be directly responsible for seventy-two deaths but would indirectly cause many more. Thousands of homes would be destroyed, millions would be left without power, and the billions in damage would make it second in expense to only Katrina.

—

In the wake of Sandy the prophet and I will return to the drowned city. Trailing his robes behind him, he will point his wooden staff at the places where the waters rose, where the subway steps became waterfalls, the cross-streets turned inlets. We will walk to where the great gaping hole in the ground once was, where the buildings now rise, and he will gesture toward the river, explaining how it backed up from winds and tide, how the full moon affected this most modern of places. Three years before, when he pointed to those same spots and told me what was coming, I only half believed him. Now I believe, along with everyone else. We have seen it with our own eyes.

This is how I pitched it to magazines.

Outside magazine bit. They offered to pay for my trip up the coast with Orrin. The plan was to drive up the Outer Banks and then along the coast up to New Jersey and New York in the late winter of 2013, several months after Sandy. We'd follow the hurricane's path and do a postmortem on the storm, report on the places it had hit the hardest.

On the morning of Sunday, February 27, 2013, I made the two-and-a-half-hour trip to Orrin's house in Durham.

By then we knew each other's habits well. When I called the day before to ask if I could show up at his house at six on a Sunday morning, he said fine, he'd have a pot brewing.

When I reached Orrin's house I thankfully took a cup and, while he threw some last-minute things in his suitcase, I chatted with his wife, Sharlene, in their kitchen. The house was well out of town, on a hillside on many acres of woods above a small highway, and Orrin and Sharlene had lived there for forty-six years. They had raised their five children there. All of those children—and Sharlene—had eventually coauthored books with Orrin. (Looking back, I'm sad that this would be my one and only conversation with Sharlene, but I remember her as sharp and funny, even in those predawn hours. She would die two years later.)

Orrin said goodbye to his wife and we pushed off in my rental car and headed into Durham to pick up Jeremy, the photographer the magazine had hired. Jeremy seemed like a nice guy, young and clearly accomplished: both Orrin and I had seen his recent spread of pictures on the *New York Times* sports pages, which featured the former Duke player Jason Williams, who had almost died in a motorcycle crash.

That first day we drove out along the Outer Banks and Jeremy snapped pictures of a sight that had now become familiar to me: battered houses on the low-tide sand. We spent the night in Kitty Hawk and the next morning made our way up into Norfolk, Virginia. We were headed to Old Dominion University to meet Dr. Larry Atkinson, a professor of oceanography and an old colleague of Orrin's. These were the pre–smart phone days and all we had was a scribbled address on a piece of paper. Orrin and I, left to our own devices, would probably have been a good hour or two late. But we were in luck: we had a young person with us. Jeremy, crammed in the back seat with his photo equipment, pulled out the equivalent of a magic compass. He took the scrap of paper Orrin handed him and typed the address into his somewhat-smart phone and soon that phone was talking to us. He repeated the directions the phone gave him, shouting them out: "Turn right here." "Go left at the third light." Orrin chuckled each time the directions were called, and laughed loudest when we finally pulled up at our destination right on time.

We sat in a small school dining area and Larry Atkinson outlined why Norfolk had become the poster child of both sea level rise and a city's attempts to adapt to it. Sandy had barely grazed Norfolk but still left it under three feet of water. The town was used to flooding: through an unlucky combination of low elevation, subsiding land, and rising sea level, the downtown streets were often underwater.

"These storms are coming more frequently," he said. "Ten-year storms are now coming every two years."

Amazingly, the Virginia legislature had recently followed North Carolina's lead of denying sea level rise, with one state senator calling it a "left-wing term" while suggesting that it be replaced by the phrase "recurrent flooding."

But in Norfolk at least there seemed to be little bickering over the reality of the rising sea. Larry had been impressed by Norfolk's mayor, who, rather than debating sea level rise, had simply "jumped over it" as a political issue, and had gotten right down to the practical business of adapting. Larry quoted the mayor as saying: "Our town is sinking and the sea is rising. We need to find areas of retreat."

Both the mayor and the scientific community knew that the problem, whatever you chose to call it, had to be dealt with. And in this they had an unlikely ally: the American military.

"What is really going to save this area is national security," Larry told us.

Right down the street was the biggest naval base in the world, and while others were debating whether or not sea level rise was real they had gone ahead and decided to raise their docks. After all, the military hinges on readiness and if the oceans are really rising—as almost every reputable scientist, even back then, said they were—then they had better be ready. It wasn't a conservative or liberal issue but a practical one, and ironically the Navy was one of the few groups that didn't indulge in the endless military metaphors of attack and defense.

"They're way ahead on this," was how Larry put it.

That's because the military is in the habit of listening to experts, and in the case of sea level it is the scientists, not the legislators, who are the experts. It's a simple enough premise. If scientists study things and then predict something will happen, and then that thing *does* happen, it is perhaps wise to pay attention to what those same scientists are saying will happen next. Oddly enough, quite a few people didn't and still don't put stock in this notion.

We were not the only visitors to Norfolk that afternoon: President Obama was touring the shipyard. Perhaps for that reason traffic was bad, and the dumping rain didn't help. Despite the weather, we decided to take a quick tour of downtown, which was surrounded by a seawall to hold back water during storms and the highest tides. Sometimes this worked, but not always.

"It's not unusual for us to drive through seawater to get to work," Larry told us before we said goodbye.

There was something medieval about the seawall that encircled the town, particularly the massive retractable gate that could be closed during storms. More impressive to my eye was the battleship *Wisconsin*, an enormous World War II relic that was now chained to a dock, not a hundred yards from the buildings of downtown. It weighed forty-five thousand tons and was a hundred feet wide but floated, according to the sign next to it, in only two to four feet of water. I imagined it breaking free of its chain and floating down Main Street in a great storm surge.

We were soaked by the time we got back in the car. It poured the whole way as we drove across the Bay Bridge, and I decided it was best not to voice my worry that, due to the horrid visibility, we might go plunging off into the bay. The rain got even worse as we made our way up the Delmarva Peninsula. We stopped at the Great Machipongo Clam Shack, a place I still liked though everyone said it used to be better. No matter. We were happy to be out of the car. Orrin devoured a bowl of crab soup and a plate of scallops. Jeremy and I both ordered a dozen oysters, and after slurping them down, headed back out.

The car was buffeted by the winds and the rain lashed. We finally made it to Ocean City, Maryland, an example of coastal development almost as rabidly overreaching as that we had seen on the Outer Banks.

After we had settled in a hotel, Orrin came down and joined me for our nightly beer.

"Hey, this trip is really fun," he said, which made me happy.

Then Orrin told me about a funny interaction he'd had with Jeremy earlier. Jeremy had mentioned that his father also worked for Duke, but for some reason Orrin thought he might be deceased.

"Is your father alive?" Orrin asked.

"Yes. He's the Duke provost," Jeremy said

"Oh, then I guess he is still alive," Orrin said.

I asked Orrin if he knew the provost.

"I sure do. Sharlene was at a planning board meeting in town one time and the provost was there. There was a heated debate about the Duke Forest, and he and Sharlene took opposite sides. All of a sudden he exploded and out of nowhere he yelled: 'Orrin Pilkey never goes to faculty meetings. He doesn't prepare. He is terrible on committees.' It was all true, of course. But the rest of the planning board just sat there in shock over his outburst. He'd apparently bottled that up for a long time."

"And now his son is stuck in a car with you," I said.

"Yup, poor kid."

The next morning I got up early and stood out on my hotel balcony in my boxers staring down at the cold, empty, winter beach. There was an element of ghost town to all the barrier islands we were driving through, and Ocean City was no different. Empty hotels, empty restaurants, empty miniature golf courses. With the roar of the winter ocean providing the soundtrack, it did not take an apocalyptic imagination to see that this place was a nightmare waiting to happen.

For miles the beach had an industrial feel, the hotels, patios, and boardwalk, and a concrete wall running along the edge of the sand, and then a forty-yard slope of open sand to where the gulls picked at the edge of the water. The beach was well-graded, with tire tracks running down it. But if it seemed somewhat dull and predictable, the sun rising over the water was not, and soon an orange glow filled my hotel room.

My friend Dylan McNamara, who is head of the physics department where I teach and focuses much of his work on climate, grew up here in Ocean City. When I asked him my forty-two-year question, he replied cautiously about what the possible sea level rise would be and then painted the picture of his hometown in 2063:

> I'd predict towns like Ocean City will look a bit like Detroit after the collapse of the auto industry and/or like Atlantic City on a bleak winter evening. There will still be many structures—hotels, restaurants, miniature golf courses—but they will mostly be abandoned or in disrepair. Some folks will still vacation there but they will do so at a very low cost. It will be a dying city on the sea.

He continued:

> I think it will be impossible to get homeowner's

insurance for any residence in Ocean City by 2063. The property value will have dropped since repair costs from storms, floods, etc. will have to be covered by home-owners and that cost will reduce the value of homes. On top of this, the tourist revenue will be much lower for a variety of reasons and thus the amenity value of coastal homes will have dropped. All of this will significantly reduce property value. This will mean that beach nour-ishment will no longer be paid for by state and federal governments. The cost-to-benefit ratio will not justify using taxpayer dollars from outside the local area. The local tax base will not be large enough to cover the cost by itself. So by 2063, with no way to fund beach nour-ishment, the beach will be nearly nonexistent in Ocean City other than maybe at low tide. The environmental disaster that looms beyond 2063 from rotting infrastruc-ture taken by an even higher sea will be the hot-button political issue. Who will pay for the cleanup and how?

We ate breakfast at the hotel and headed north toward Lewes, Delaware. We drove through a strip-mall landscape with blocky high-rise hotels on our right, toward the ocean, and the usual mix of McDonald's and Hooters and other chains on our left, but at one point we passed an open stretch of water, sand, and low gnarled trees.

"Why do they have all these trees here?" Orrin asked. "Don't they know this would be a nice place for some condos?"

In Lewes we took the ferry across Delaware Bay. At one point Sandy, after it was supposed to hit my hometown and then the Outer Banks, was supposed to come driving up into the bay we were now crossing. But if we had learned one thing over the last couple of years it was that hurricanes don't always do what they are supposed to. It was funny but both sides in the coastal debate liked to use the uncertainty of storms to bolster their arguments.

Those like Orrin who argue for moving back from the coast can say that a storm can strike at any time and any place and from any direction. Those on the other side can say the same and argue that no place was safe. And if it's all a crapshoot anyway, why not rebuild?

As we approached the other shore, buffleheads, small ducks with white cheeks, bobbed on the surf. The ferry pulled in and docked and we drove off, heading to the beach that fronts the town of Cape May. Bruised clouds muscled overhead as we climbed out of the car, and the wind strafed the sand. It was a classic cold, choppy white-capped winter beach day, the foam kicking up, though surfers in wetsuits were still out there braving it. Orrin pointed out that the sand was porous, "bubbly," he called it, the kind that was bad for oil or any other kind of spill since foreign material would sink in deep.

Cape May had been one of this country's very first seaside resorts, and by the early 1800s was already luring tourists to its shores. In fact, New Jersey as a whole was on the vanguard of beach tourism, drawing New Yorkers and Philadelphians to its beaches, just as the North Shore and Cape Cod later drew Bostonians. It was here that people first noticed that the beaches were moving shoreward, that they were, as residents perceived it, "losing their sand." And so it was here that they unintentionally embarked on what Orrin would call "a giant science experiment" by building hundreds of seawalls, jetties, and groins.

Sand *flows* along the shore. It flows downdrift, a sand river that can move either north or south, depending on the geographic conditions and circumstances. It is something that most of us don't notice, but the residents of Jersey Shore towns picked up on this fact pretty quickly. They began to erect groins on their flanks that would capture the sand before it migrated down to their neighbors.

To a trained eye, this beach still provides a classic example of how groins work. A series of rocks juts out perpendicularly from the beach. Many beach communities call these lines of rocks "jetties," though technically they are "groins" and the beach is a "groin field." Groins, though admittedly not a beautiful word, is the right one since jetties actually refer to rock barriers that run along the beach, not perpendicular to it, or to piled rocks that enclose a harbor. (Though where I grew up everyone called the rocks that jutted out into the harbor a jetty.) The general effect of a groin is to gather sand on its updrift side and cut away sand from its downdrift. These structures give, but they also take. In short, the updrift folks were screwing those downdrift from them. These groins even had their own apt name. They were called "spite groins."

Cape May provides a beautiful illustration of how groins work. On the updrift side of the groins the sand is built up in a large white-backed hump; on the downdrift side it is carved away almost completely. We walked out to one groin where, on the updrift side, sand reached halfway out the forty-foot groin, while on the downdrift water splashed on the very first shore-ward rocks.

We climbed back in the car and continued our drive north. When we passed the Parkway exit for Corson's Inlet, I told Orrin that I'd spent some time hiking along the inlet. It was one of my favorite spots on the drive between Cape Cod and Wilmington. One day I'd wandered through the dunes and underbrush and watched thousands of tree swallows in the midst of their fall migration. Fifty years before my own walk, the poet A. R. Ammons had watched the same phenomenon with an earlier generation of swallows. He noted that the exact pattern of that particular day could never be repeated. This was true, but I had witnessed some of the same fundamentals that he had recorded in his brilliant poem about the experience, "Corson's Inlet": the cyclonic lifting of the birds, the movement of the whole as one, the sudden scattering of the collective into individual flight.

I mentioned the poem to Orrin. Ammons, who grew up in

North Carolina not far from Wilmington, in the small town of Whiteville, spent his adult years living in the North. He is, for my money, this country's best coastal poet. While traveling around the country's beaches over the last few years, I have often carried with me Orrin's *How to Read a North Carolina Beach.* But I've also carried another book, a slim volume that has been as important as any field guide in helping me learn about the coasts. That volume is Ammons's *Selected Poems.*

His poems about the Jersey Shore, particularly the southern Jersey Shore, interest me most. They seem to articulate with incredible exactitude the pulsations and movements, the uncertainty and precariousness, of life on the coast.

"Firm ground is not available ground," Ammons wrote in the poem "Dunes," a line that has become my mantra as I study the shore. He is referring to beach grass, but might be talking about anything that tries to eke out a life near the water. Study the way a sanderling darts in and out as the waves crash and recede, hunting for mole crabs in a place in between water and land. Or the way a pitch pine grows, gnarled and small, leafing to leeward, adapting to constant coastal winds. Or the way that human beings try to control an uncontrollable landscape.

Then there is his poem "Saliences." It begins:

> *Consistencies rise*
> *And ride*
> *The mind down*
> *Hard routes*
> *Walled*
> *With no outlet and so*
> *To open a variable geography,*
> *Proliferate*
> *Possibility, here*
> *is this dune fest*
> *releasing*

mind feeding out,
  gathering clusters, fields of order in disorder

To be honest, I don't always understand everything Ammons is trying to say. But I understand the push for openness to the actual, for not forcing consistencies—and walls—on fluid things. On a philosophical level, this jibes with how life actually feels to me, particularly life near the water. On a practical level, it also seems to match the way we should try to live on the shore.

As an advocate, there is a consistency required in Orrin's thinking—at least his public declamations about that thinking—that is not required in my line of work. That is, there is something like a wall in Orrin's opposition to walls. At the same time, Orrin's uneasiness with scientific models stems from something like my own uneasiness with those who plaster grand theories over messy reality.

Ammons believed, not in maps of things, but in the thing itself, since no map can ever replicate the basically irreducible complexity of the real. He believed in "no book of laws, short of unattainable reality itself…" Natural processes have their own particular logic, and that particular logic works differently on different days, in different times, tides, and weathers. This sounds almost exactly like the reason Orrin distrusts quantitative models: they create a picture of the future based on oversimplifications of the sloppy world, without enough "ground-truthing."

Traveling the coasts, I have come to believe that we overvalue certainty in this world. Meanwhile we try to pin our own limited and overly logical manner of thought on things that will not have it. We lay down mental concrete but weeds break through. This is especially true in studying earth systems. In physics, by contrast, the rules, for the most part, are the rules. Not so in the natural world where too many complex variables are at play, sometimes in ways they may never have played before. The other day, not long before we left on our trip, I took a walk in the woods and saw a small seedling dangling from a tree. It looked like a tiny ornament, spinning wildly in the southeast breeze, and it occurred to me that the pattern it played out, the *exact* pattern of back and forth and then up and down, had likely never been played out in that exact manner in the billions of years that earth has existed. In other words, you would have a hard time predicting how that leaf would twirl, or the future of that one acre of woods, let alone the coming climate of planet earth. It is the same with any large process of the planet. Each day brings something different that adds to the equation. Or as Ammons puts it: "A new walk is a new walk."

It's an ancient war, one that's been going on forever, between those who want to clean things up and those who accept the mess. The theorists hate sloppy life. They hate anything unclean. They don't like the chaos that's been handed down to them and want to neaten it up. They are the room-cleaners, the orderers, valuing certainty over creativity and fluidity. I know these people. I have butted heads with them all my life. As has Orrin. They simply don't understand that in nature sloppiness is next to godliness.

As we drove north, I decided to attempt a grand bridging of the poetic and scientific. In "Corson's Inlet," A. R. Ammons wrote one of my favorite lines about how the natural world works, and I now repeated that line for Orrin: "No humbling of reality to precept."

Orrin liked this phrase but challenged me to say in plain English what it meant.

"It means that reality is messy and rules don't necessarily apply," I said. "That any rules about a place should evolve from a particular place."

Orrin considered this for a minute and then gave it a "Hmmph!"

—

We had traveled through many of the places where the storm

was supposed to hit. Now at last we drove across a bridge and out onto one of the places where it really did. Long Beach Island, halfway up the Jersey Shore, sat just north of where Sandy's eye passed.

Long Beach is eighteen miles long and at the time had a year-round population of twenty thousand. Parts of the current island were formed by dredging the backside marsh, and it showed. Five years before I had taken a tour of this same island, not with Orrin but with a former student of his, a geologist and coastal planner named Sue Halsey. Sue had pointed out how poorly designed the streets were, running straight and uninterrupted from the island's ocean side to sound side, so flat that you could roll a bowling ball across the entire island. She predicted that, when a storm hit, the waters would rush in and turn those streets into inlets.

Which is exactly what happened on the night of October 29, 2012. The winds came in out of the east then cranked around to the north and northeast, turning with the storm, gusting to eighty miles per hour and sustained at sixty, for about three hours. Not the strongest of winds, really—double that is possible next time—but Sandy made up in size what it lacked in speed, with storm-force winds five hundred miles offshore that meant for a huge wind fetch, *fetch* being the amount of distance wind can travel unimpeded. "Straight from Casablanca," as Orrin put it.

Then there were the most primal of factors: the storm hit at high tide on the night of the full moon. Water and sand churned down the streets. Houses were bullied off their foundations, left in splinters, knocked clean across the island and out into Barnegat Bay.

It doesn't take a hurricane to flood Long Beach. The unnamed nor'easter we had driven through the night before on the way to Ocean City had done the trick, and as we drove we splashed through the flooded streets. We walked down to the beach and

Orrin admired the quality of the sand. The storm had blown through and left a beautiful, bracing day, the wind blowing east, pushing the foam of the whitecaps back out as the waves came in. We studied the artificial dunes that had protected the houses. The trouble was they weren't consistent. Before Sandy many of the homeowners hadn't signed the easement that allowed the dunes to be built in front of their homes, complaining that it ruined their view. During the storm the ocean wasted no time finding those openings and barreling through. All it needed was a gap. The homeowners' views that night were of water charging at them like a freight train.

Walking back from the beach, we bumped into Dennis and Sean Cleary, father and son, who lived right on 82nd Street. Like everyone else they were still digging out, four months later.

"There was a ten-inch-thick concrete slab under this entire house," Dennis said. "Now it's completely gone."

"Was it pulled out into the ocean or swept through town?" I asked.

"One of those," he said with a smile.

Dennis and Sean had been on the damage assessment team and had stayed the night of the storm.

"The next day we drove around and saw no houses where houses used to be," Dennis said. "But there would be this hissing noise. The gas lines hadn't been shut off yet and were spewing out gas."

He told us that the storm couldn't have had a more perfect name. They found the streets covered with four to five feet of sand, through which they eventually dug single-lane roads, the sand piled up in banks on the roadside like snow. In those cross-cuts you could see geologic layers of mulch, sand, wood, and stone.

"This guy didn't sign the easement," Sean said, pointing at a neighbor's crumpled house. You could almost argue that this neighbor got what he deserved, but it wasn't really worth looking for justice in the storm's wake: the man's destroyed house, for instance, had come unmoored and taken out the house behind it, which belonged to a friend of the Clearys.

After the storm there was a great rush to sign the easements, but there was still one neighbor who was holding out.

"A monumentally selfish act," Orrin said. "How idiotic to be worried about their *views*."

After we said goodbye and headed north again, Orrin shook his head.

"It's hopeless anyway," he said.

I asked him what he meant.

"An island that's as flat as a pool table with only a seven-foot elevation straight across."

I knew what he was driving at. If the sea rises anywhere close to the same height as the island's current elevation, dunes won't help. Even with a more modest rise, the island didn't stand much of a chance, though as usual this fact hadn't given pause to the many who were already rebuilding. Included in the sixty billion dollars in the government aid for Sandy relief, rushed through in a patriotic fervor, were millions of dollars for thousands of houses just like these, houses in the worst of the flood zones. It was, as Orrin had said, like giving money to people to help them rebuild on a train track, and then telling them to just keep their fingers crossed and hope the train doesn't come again.

"Yup, this place is hopeless," he repeated, this time with more enthusiasm.

"That's not much of a political platform," I said.

He smiled. "Of course it's not to be said out loud around actual people."

—

The next morning we toured the town of Mantoloking, which, in terms of pure destruction, was as close as Sandy had to a ground zero. Our tour guides were Matt Totin and his mother, Kathy, who lived just to the north in Point Pleasant Beach. As it turned out, Kathy, who was a former councilwoman from central Jersey and a lifelong resident of Point Pleasant, was a fount of local knowledge, as was Matt, a self-described "weather maven" who had studied Orrin's books in college and who could casually drop phrases like "downdrift erosion" and "wave action" into his conversation.

We met them in Point Pleasant Beach at their small, army barracks-style home, on a side street two hundred feet from the beach. The flooding had stopped one house short of them on the ocean side, and they had spent weeks digging out, dumping all the sand in the street where it was moved by state emergency workers with backhoes and trucks.

"The power was out for weeks," Kathy told us. "It was primitive. Then finally I had everything fixed but the cable. One day a Cablevision truck pulled up on the streets. I was excited but then the Cablevision guy went to the back of the truck and took out his surfboard."

"The waves were really great after the storm," added Matt.

Matt, also a surfer, gave us the local skinny on Sandy.

"The high tide was everything here," he said. "Eleven feet was our previous high tide record, but this one was thirteen and a half. The storm hit at high tide and on the full moon. Without the full moon this would have been the '92 storm all over again. The moon is what accounted for the extra two feet."

Matt was a virtual glossary of coastal terms. Some that I knew, some I didn't. I stopped him when he said that the ocean had been "stacked up" during Sandy, and asked him what that meant.

"It's a phrase you'll hear from surfers or from commercial fishermen who survived a rough trip," he said. "They are explaining a condition where waves no longer come in sets, but in a seemingly endless march of ever-bigger peaks…'stacked' to the horizon. I've seen this in the rarest of storms, when the tide is relentless, as high as you've ever seen it, completely defying you to come near the beach."

Matt explained exactly why Sandy did so much damage. The difference between the average storms and the damaging storms is not a matter of sheer strength so much as forward motion. As a storm sits off the coast, a gale-strength northeast wind blowing across a great fetch has the ability to push tremendous amounts of water toward the coastline. The longer period of time a specific storm sits offshore, the greater amount of water is pushed toward the beaches, bays, and inland estuaries. Until that onshore wind subsides, water cannot recede back to the sea. If a fiercely powerful storm travels up the coast at a quick rate, there can be little damage because the storm might be here and gone in less than eight to ten hours. However, if a moderately strong storm moves up the coastline slowly or is blocked by high pressure over Nova Scotia as this one was, and sits offshore for forty-eight hours or more, the storm can push the water toward the coast over the period of a few tide cycles. As a result, the bays and beaches take a beating from an increasingly rising tide.

We walked down to the broad beach behind Matt's house. It was a dangerous beach, he told us, often dragging swimmers out to sea. One summer Matt had found part of a dead female body in the wrack line. Almost as shocking, to us at least, was that the beach was privately owned, by one family, and that to put your blanket down on that private sand you have to buy a ticket and wear badges.

"You can't own a beach," Orrin said. "That's like owning a grizzly bear."

But one local family, and now another, had owned this particular grizzly bear for thirty-seven years, and had charged people to use it during that entire time.

Matt gave us a tour of the neighborhood, and we got our first glimpse of what I've come to think of as the North-Carolinazation of Jersey. Many of the houses that were flooded were now being raised up, rebuilt on stilts with no first floor, just like back on the Outer Banks. This was not due to the good sense of his neighbors, Matt explained, but because the zoning has changed.

"The insurers and banks learned their lesson from Katrina. They gave people the money and people upped and left. That's one of the problems now. The banks are holding the money to make sure people have plans. And the plans must include raising the houses."

Which could cost over a hundred grand, on top of substantial boosts in flood insurance, not to mention the rebuilding itself. Orrin's longtime hope, so far unfulfilled, was that those concerned with the bottom line would become his allies. That if coastal geology wasn't going to stop people from rebuilding, then raw economics might.

"One guy up in Mantoloking had his house up on stilts already," Matt told us. "Everyone bitched and moaned about the way it looked. But his house survived."

The water had flowed through it, of course. This may be the central lesson of coastal living: don't block the flow. It occurred to me, not for the first time, that anyone concerned with living on the coast should sit down with a big bag of popcorn and watch reruns of *Kung Fu*.

Before we drove down to Mantoloking we were joined by Sue Halsey, whom I had last seen five years before when we toured Long Beach Island. Her license plate—"4Dune"—announced her beach politics, and she, opinionated and brash, announced herself. Sue was short enough that even Orrin towered over her. We all climbed into Kathy's SUV and drove down to Mantoloking, and right away I could see that Kathy and Sue would be jockeying for the role of chief tour guide. Both Kathy, the local, and Sue, the expert, wanted to narrate the story of what we were seeing, and for a while it was a little like being in a Robert Altman film, their talk overlapping.

All talking stopped when we reached Mantoloking. Four months after the storm and the place still looked like it had just survived aerial bombing. Here you could see the future. The signs of devastation made those on Long Beach look mild. Crumpled houses, houses floating out in the bay, houses cracked in half with their innards—TVs, rugs, lamps, books, and in one even a comfortable-looking easy chair—revealed, like some giant's diorama of a human habitation. Four months had passed and the homes remained deserted, the power still off. Huge mansions lay splintered, and, on the bay side, formerly deepwater docks now floated not on the water but atop sand and debris.

We parked in the middle of town, which had been wiped clean. Here the water had breached on the ocean side and torn through to Barnegat Bay, creating an inlet where none had been and sweeping everything in front of it out into the bay. A corrugated steel wall now served as a bulkhead where the water had rushed through.

State police watched from every corner, keeping drivers from stopping and gawking, and, of course, from looting, which had been a problem early on. We procured a press pass and walked down the devastated streets. The first house we came to slumped into the bay, half on land and half on water, and the second lacked a front wall: we peeked in at a dining room where everything was shattered except for a still-hanging chandelier. Out in the middle of the bay a house floated, flying a flag like a lone sail and listing to starboard.

Kathy told us that she had cried when she first saw this. While Orrin wasn't crying, he wasn't exactly gloating either. In fact, overall, Orrin had more sympathy for these residents than those of the Outer Banks, since here they had no historical reason to expect such a storm. Implied in this was the fact that he wouldn't be as sympathetic next time.

Matt pointed out that there was a certain element of randomness to hurricanes.

"Look at this house," he said. "It's untouched. The grill still sitting in the backyard. And now look at this one."

He pointed to the house next door, or what was left of it. It appeared to have been chopped up into tiny pieces by a team of manic lumberjacks. He said that he came to this beach to surf as a kid but that most of the landscape was now unrecognizable, having been redesigned and reordered by the storm.

"The rich got hit the hardest," Orrin said. "The best houses got it worst."

In this case it was true. Those who chose to live right next to the sea got to learn a little about the consequences of doing so.

I wandered off alone down to the beach. I'm generally against military metaphors, but maybe it's impossible to talk about something like this without them. Tattered American flags flew from the top of crumpled buildings, and without thinking I found myself using words like *ground zero*, *attack*, and *battle*. I even briefly pitied the poor politicians who opposed the billion-dollar boondoggle of relief that passed right after the storm, and who finally caved even though the relief bill contained money for fisheries in Alaska and dredging along the Mississippi. No matter. You didn't want to get caught using words like *retreat* right after Sandy.

But to temper this I reminded myself that most of these houses were second homes, vacation homes. The rich got it worst, as Orrin said, at least here. And in the end the military metaphors simply don't wash. The ocean isn't really the enemy: no one wants to "defeat" it (and no one with any sense thinks they can). In fact, people moved here precisely because they like the ocean, its wildness and beauty and uncertainty. True, they like it in controllable doses, certainly not in the dose they got it in. But the larger point is that the sea is not Grant's army laying siege to Vicksburg. It's an elemental force that has been doing the same steady work for eons. Is it unpatriotic to suggest that respecting that primal force, even getting out of its way, is a better idea than fighting it?

Everything is ephemeral, Master Po might tell Kwai Chang Caine, nothing lasts. We know this on some level. These houses are children's castles in the sand waiting for the next wave to roll in. And yet...and yet most of us don't—and maybe can't—live that way. We bolster that castle. Build our lives around it. Maybe put in a two-car garage or a pool out back. Treat that castle like it will be here forever.

The trouble is that during the thousands of years of human existence on earth, living by the sea has always been a gamble, carrying with it the decent chance of being wiped out. That's why the shore was often merely a migratory destination, a place to move to during the milder months and away from during the stormy ones. Humans built shacks, knowing they might be wiped out. But now we have chosen to build castles instead, this seemingly insatiable drive for more coming from the same root drive that leads us to burn fuel like drunken pirates, the very thing that is leading to the heated ocean and rising seas. What Orrin was saying was simple: this is idiocy, you will be wiped out. And, Sue Halsey might add, if you are going to defend yourself, at least consider softer defenses, natural defenses, like dunes.

But even after Sandy few people were comfortable with the soft, the squishy. We like things hard, we like our walls, our rocks, our concrete, our sandbags, our war metaphors. That is how we will fight back against the water. That is how we—flags waving—will finally defeat the sea!

My thoughts on beach wanderings were interrupted by a cop, who came bombing toward me on an ATV. He seemed to be having fun, but he wiped the smile off his face to warn me that it wasn't safe down here. When I pointed to the surfers to the north, he said: "They're morons. The water is full of metal. Lots of rebar."

After I returned to the group, we climbed back in the car and drove south, toward the land of Snooki and the famous apocalyptic-looking roller coaster that stood stranded out in the water like a giant metal pretzel. The competition for chief tour guide started right up again, and Sue kept interrupting until it got to the point where I wouldn't have been surprised if Kathy jumped over into the back seat and tried to gag her.

"The inlet wants to be an inlet," Kathy said, pointing at the steel barricade in the middle of town. "You can tell that it wants to be water. For all that they've fixed the breaks you can't tell the water what to be."

Kathy's words were more poetic, but Sue could be impressive.

In her role as senior geologist for the New Jersey Department of Environmental Protection, she was known as "Dr. Dune." One part of her job was to tell people living in these beach towns that a hurricane was going to wipe them out and that water would rush up their streets. Then she would tell them exactly what they needed to do to defend themselves. People didn't like this news very much and weren't crazy about Sue either. She didn't care. She was an acolyte of Orrin's, and if possible had even more of a taste for controversy than her old teacher.

We now entered Lavallette, a town that was a case in point, though I doubted that anyone there knew that we had, crammed between Matt and me in our back seat, the town's savior.

"I made them create dunes here after the storm in '84," Sue said. "And they hated me. I had death threats. But during Sandy those dunes took the brunt. You know it's like a bumper on a car."

"Lavallette did really well," Matt confirmed.

"What do you mean, 'you made them?'" I asked.

She told us that after the '84 storm, which took out the boardwalk, she headed the Department of Environmental Protection team called in to mitigate. Sue informed the locals that if they wanted the money to rebuild their boardwalk they would have to put in dunes first.

"Oh my God, they said I was the Wicked Witch of the West, and they said, 'What the hell does she know?' And I said, 'It's simple. You do it or you don't get your FEMA money. I'm the one writing the report.'"

The first dunes were pathetic scraggly things but then they grew, as dunes will. They helped defend Lavallette during the '91 storm, though the '92 storm took them out. It didn't matter. This time, without threats or cajoling, the town went ahead and built the dunes back up. By then they were believers.

"Now *dune* is a good word here," Sue said. "It used to be a bad word."

Obviously Sue Halsey was not above a little I-told-you-so. And why not? She had, after all, told them so. And if she was a know-it-all, the fact was she did know it all, or pretty close: FEMA rules, science, local politics, history. If every town on the coast had been forced to listen to Sue, Sandy wouldn't have done nearly as much damage. A short drive in either direction confirmed that the neighboring towns didn't fare nearly as well, water roaring down their streets. In fact, they might now be wishing that they once had someone like Sue to hate.

—

There are fighters like Orrin and Sue. And then there are the rest of us.

Before we headed into the city, we needed to make a detour. I had been asked to give a talk at Drew University, which was a couple of hours inland and north of our planned trip. I suggested to Orrin and Jeremy that they get a hotel on the shore; they had just spent four days in the car with me and certainly didn't need to listen to me pontificating from behind a podium. But Orrin insisted: "I *want* to hear you." And so we left the coast for the first time during our journey. Once again we relied on Jeremy's magic phone to guide us to the university right on time. We just made it, and were greeted by my friend Patrick, a poet and a professor at Drew.

I gave my talk to a good crowd of students. The size of the crowd said less about me than about Patrick. You can tell a lot about a school by the way they host a talk, and one of the best things a school can do is require the attendance of students, giving us hapless speakers the illusion of popularity.

During the talk, I made fun of the tendency of most environmentalists to invoke the apocalypse. Back then I might have been inclined to mock the book you are holding now. I read from an older book of mine: "I suppose if I really wanted to make it big I would start spreading the word of doom and intoning the phrase 'global warming' over and over, hitting my audiences with it like a big stick."

There was laughter, which pleased me. But when the time for the Q and A came, it wasn't the poor, captive students who raised their hands—they had done their time already, thank you very much. Instead it was the bearded emeritus professor from Duke.

He held his hand up like a schoolboy until I called on him. It turned out, not surprisingly, that his was more of a comment than a question.

"If all the ice melts—and it certainly will eventually—then the water will rise two hundred feet and we will all expire. But the earth will go on. Humans, after all, are just one species on earth. There have been mass extinctions before."

And what was I, the esteemed speaker, supposed to say to that?

"Good point," I managed.

"We can't be afraid to be ourselves," is an expression I've always liked. Orrin certainly wasn't afraid of being Orrin. He had a sense of humor, I knew that. But he also had his main theme and he was sticking to it. He had a flag and he would wave that flag at every chance.

Over the years I have come to believe in his flag. Jokes are great, equivocations part of life, and yes, there is more to being alive than advocacy. But this dark thing is coming toward us and it is coming fast. Why do we do everything we can to ignore it?

Orrin's roots are in science. Mine are as a kind of nature romantic. But over the years, as I have traveled through this country and beyond, a darker reality has begun to permeate my romantic journeys in nature. The snake had been there in the garden from the beginning, but in the early days it was a more local and specific snake. I railed against developers who tore down trees to put up trophy homes, industries that belched foul smoke into the air, chemical companies that introduced poisons into the very chain of life. In an early book, *My Green Manifesto*, I wrote: "If I had to give one piece of practical advice it would be this: Find something that you love that they're fucking with and then fight for it." But I was relatively slow to accept the fight at the climate level, and back then I didn't quite accept that what they were fucking with was the whole wide world.

As I have grown older, and evolved from a moment-seeking nature romantic into a kind of reporter, I have witnessed global warming (to call it by its less-watered-down original name) change from an obscure idea that few understood into a theory that does almost as good a job describing the reality of our present world as Darwin did describing why animals evolve. I have now seen the climate crisis with my own eyes. Seen places gutted by the search for fossil fuels. Seen Vernal, Utah, fracked and Barataria Bay, Louisiana, covered in a film of oil, not to mention the water rising in my own backyard. This is the real reality show today. This is how we live. The price tag for our nonstop spending spree.

I have also seen our environmental battleground shift from the local to the global, and watched our crowded world be increasingly ravaged by drought, heat waves, fire, massive storms, species extinction, and human displacement. And I have been increasingly nagged by the question, the forever question, *What is to be done?*

I, like so many of us, have done little. Yes, I have written books and articles and given talks. But that is not enough. I have sat on the sidelines politically, or at best been a cheerleader for the green team. The writers I admire most these days are like Orrin, going about their business of making art while also fighting the way the world is heading. This second part is often dull and difficult and opens one to criticism and censure. It is here that I, like so many of us, have fallen short.

I have been slow to grow a conscience. Maybe you have, too.

Or maybe I have a conscience that nags at me, but I try to ignore.

Why? Why haven't I changed? Why don't I fight instead of just write?

I honestly don't know. Maybe because it is hard to change habits, maybe because it is easy to do what we've always done, to live the way we have always lived.

Some environmental thinkers say that to focus on individual reform is not the point: without regulation and corporate reform and laws this means little.

And they are right. Alone we can do little. Change can only happen if we change the laws of the land with regard to fossil fuel use. *Really* change them. That is taking nothing away from those individuals who are trying to live environmentally sound lives. But that ain't going to cut it. The rules need to change. A fairly ridiculous example of the effectiveness of regulation springs to mind. I think back about how I, as a teenager and young adult, railed against the idea that we should be forced to wear seatbelts. Back then only wimps wore seatbelts. What an infringement on our freedom! How preposterous that seems now, but also strangely hopeful in that it shows how quickly we can change our habits when required to. So yes, if we change the laws people are likely to change, and to quickly forget they have changed.

But while this may be true, it still does not get me off the hook.

I have now written myself into a corner. Like so many of us, I go about most of my life essentially ignoring what is happening to our world and what we might do to stop it. Like so many of us, my focus is on all the crap I've got to do today instead of on what is heading toward us in days to come. And if I, frequent witness to the way that fossil fuels have destroyed our world, still pump my gas and watch my shows and whistle in the face of doom, what chance do we have?

My instinct here is, like most essayists, to tie this up and give us a nice ending.

But I am not going to do that. Instead I am going to end here, in the same confused place that many of us find ourselves in this dark time.

Dangling.

—

Our Sandy procession continued north.

Sand and empty coffee cups littered the floor of our rental car, which had taken on the feeling of a dark, dank cave. The three of us, Orrin, Jeremy the photographer, and me had made our way up the coast, stopping at Asbury Park where the boardwalk was torn up and where two men, one Black, one white, seemed to have been given the responsibility of rebuilding it entirely by themselves. They took a break from hammering to light cigarettes and talk to us.

"People stayed," the Black guy told us. "They just said 'nah' and then had to be rescued in pontoon boats. I stayed and it stopped a hundred feet from my house. All night long it sounded like a train."

It was a spectacular cold beach day and we walked under blue skies and flying gulls, down through the great hulking arcade buildings that listed like wrecked ships on the low-tide sand. Grackles and mourning doves nested in the girders that ran below the buildings.

From Asbury Park we drove on to Sea Bright, a town that Orrin had helped make famous in the coastal science community as the most egregious example of the danger and destructiveness of seawalls. The whole drive through Sea Bright, during which we were never more than fifty yards away from the beach, we never once saw the ocean, just a twelve-foot-high wall off to our right. It was an entirely medieval scheme, to wall off your town from the enemy, the epitome of the military, anti–*Kung Fu* philosophy of the Corps of Engineers. As Orrin had long pointed out, there were many problems with such a wall, one being financial. After the very first storm it weathered, the cost of repairing the wall was more than the net worth of the town it had been built to protect. But worse was the way that the wall swallowed up the beach, requiring that it be constantly "nourished," which was a quaint way of saying "have millions of tons of sand periodically dumped on it after any storm." When a wall like that is breached, or topped, the water comes through full force and can't get back over, and we passed several neighborhoods that were flushed out in this way by Sandy.

"You would think people would learn," Orrin said.

But of course they hadn't. Just a week before, the Department of Environmental Protection, while saying that "seawalls are not the answer," had granted residents of Bay Head the right to build a 1,300-foot-long wall along their beach, just south of Matt and Kathy's.

"They always say this is only a temporary solution," Orrin said. "But seawalls are not temporary. They only do one thing well and that is ruin beaches."

We drove on, feeling more sympathetic when we reached the wall-less working-class town of Union Beach, where we saw a row of wrecked houses that didn't seem to be second homes. "Real people," Orrin called them. We circled out through Staten Island, which was hit hard with at least twenty-three deaths, and then over to Brooklyn and into the city through the Battery Tunnel.

Orrin noted the particular dynamics of the Hudson River, how if the tides and winds were right the river could back up over the wall at Battery Park, and indeed that is where TV viewers had last seen Jim Cantore on the night of October 29, just before the lights died.

"These people don't seem to understand they are living on an island in a time of rising seas," Orrin said. "That what happened isn't nearly as bad as what will come."

We parked at a garage near Battery Park and walked along Water Street, which fully earned its name during the storm. It was a street that, if Orrin's predictions were right, would be underwater by the end of the century—without a storm. Whether it would be or not, the more immediate danger of sea level rise is the increase in storm surge, where a small increase can lead to a large, and deadly, impact.

We talked to everyone we saw, starting with the guy at the parking garage, who told us the water was waist deep in the garage during the storm, and who took us down to the lowest level to show us a deserted, sea-encrusted car. I had a friend who lives in Manhattan who said that the lesson learned from the storm was New York's remarkable "permeability." She was amazed at how the whole place was soaked through and was up and running a few weeks later. And it was true that downtown looked pretty good at first—no obvious damage of the sort we saw on Mantoloking.

But if you looked a little closer you saw obvious signs, starting with the rectangular yellow ones, notices of evacuation and closing and condemnation, in the windows of buildings. And as we walked farther south, descending almost to sea level, Orrin pointed out the watermarks on the side of the buildings, five or six feet up. I remembered that the city's fourteen wastewater plants all sat at water's edge and had outfall pipes at sea level. During storms those outfalls quickly became infalls, sending sewage backward into the pipes, guaranteeing not just deadly disasters but smelly ones. Where we were now walking was briefly a lake during Sandy, though a lake made up not just of water but of millions of gallons of sewage and fuel, and everything else the sea swept up. We stopped and talked to a storeowner who confirmed this, describing how he watched Abercrombie and Fitch mannequins bob by in the water.

Orrin was hungry again. And understandably tired. We climbed John Street, gaining what will one day be seen as precious elevation, and decided to eat at a place called the Open Door Gastropub. We sat down, ready to feast, but then on a hunch I went up to the bar and started a conversation with a man who turned out to be the owner. "You gotta see this," John Ronaghan said, and so we abandoned our table and descended the narrow stairs to a cramped cellar where the water had flowed in. I imagined the cellar on the night of the storm: to get the jar of olives you would have to swim underwater like Shelley Winters in *The Poseidon Adventure*. When the water receded, the owners had begun to deal with the lesser plagues that struck the neighborhood: rot, mold, flood insurance.

"Everything ruined," John said, adding: "And this is nothing."

He wanted us to see his other restaurant, the Paris Café, which was down on the waterfront near the old fish market. He was due at a meeting, but the co-owner, Peter O'Connell, agreed to give us a tour of the Paris. Thirty minutes later we were walking through the restaurant. Workers had been going strong for weeks but it was still in ruins; during the storm the water had risen above the heads of where his customers usually sat.

So much effort had been put into just trying to get back to where they had been. I asked Peter the obvious question. What if it happens again?

"What if?" he echoed, thinking it over. "We're taking a shot," he said.

This was one of the more realistic sentiments I'd heard voiced during our whole trip. It contained no illusions about building walls or defending or controlling the uncontrollable. He was not saying they would fly a flag and defeat the ocean. He was saying he would roll the dice and see what came up. Not a desperate craving for certainty where none existed. But a realistic assessment. A gambler's assessment.

THE APOCALYPTIC CITY

What will New York look like when Hadley is sixty?

There are many visions for the future of the city.

Almost all of them are watery.

When Orrin and I first visited New York, the architect Adam Yarinsky had recently won a $10,000 History Channel prize for envisioning the Manhattan of the next century, a design that, based on predictions of sea level, included canals running where some streets do now. New York City as Venice.

It is not farfetched. "Many ancient cities of great fame have disappeared or are now shells of their former grandeur," writes Klaus Jacob, a geophysicist with the Earth Institute at Columbia University. "Parts of ancient Alexandria suffered from the subsidence of the Nile Delta, and earthquakes and tsunamis toppled the city's famous lighthouse, one of the 'Seven Wonders of the World.'"

Jacob has played the Orrin Pilkey role in the New York sea level debate. He is deeply skeptical about dikes and barriers, and is one of the very few people before Sandy who had really thought hard about what a major hurricane would do to New York. Unlike most of his colleagues, he had no problem envisioning, and describing, the devastation. He saw streets like rivers, flooded subways, and little chance for true evacuation, a Katrina but with millions more people. He pictured winds shaking glass and bending metal, signs flying like projectiles. Jacob's only practical solution sounded a lot like the solution that Orrin has suggested for the Outer Banks. *Get the hell out of low-lying areas.*

Some of the possible "solutions" for New York include building storm surge barriers around all the wastewater plants, zoning requirements that would incorporate climate change predictions into building requirements, barriers for subways, restrictions for coastal development, and raising infrastructure, as much as possible, above the predicted sea level rise. And then there is the big solution, the grand solution, the eventual solution: the building of dikes or other barriers in the manner of that sinking nation, the Netherlands. Years ago the Storm Surge Research Group at Stony Brook University, headed by Malcolm Bowman and engineer Douglas Hill, began mapping out a plan that includes three large barriers at the Verrazzano-Narrows, Arthur Kill, and Throgs Neck, barriers that would theoretically shield Manhattan in the manner of the Eastern Scheldt barrier that protects the Netherlands.

Of course it's not so simple.

Beyond the staggering costs, there are countless problems with constructing such barriers, starting with the debatability of their effectiveness.

"The higher the defense, the deeper the floods," Jacob has written.

—

These predictions take on a new urgency for me in December 2021.

That is when Hadley gets the email telling her that she has been accepted at New York University.

Klaus Jacob has long called lower Manhattan "the Bathtub." Hadley will be living in the Bathtub for the next four years.

Three months later Hadley and I will fly up to New York City to tour her new school. At the end of our first day my phone will tell me we have, without trying, walked twelve miles as we hiked up and down the island. It will be a great day really, seeking out vegan restaurants, lounging in Washington Square Park, enjoying the early spring weather, and ending the night watching *The Book of Mormon* on Broadway.

We are not there to contemplate the city's doom. We are there to think about her future, not the world's, and for four days I will happily keep the two in separate boxes. In fact, if I am honest, I will spend approximately zero minutes while we are there worrying about the coming storms. We are always of two minds. At least.

—

The first time Orrin Pilkey and I visited New York together was in 2010, three years before our Sandy trip. Orrin flew up from Durham while I took the ferry across the Hudson from the town of Edgewater.

That day, as the ferry streamed through the brown-gray waters of the Hudson, a great blue heron flew over the boat with deep, slow wingbeats, leading us east across the water toward the city. I watched its flight until I was distracted by the buildings that rose up in front of me. This was a view that had launched a thousand metaphors. Some had seen the vision of the city as a shining example of human progress and ascension, some as merely hubristic. Those who are environmentally inclined might have once sneered at the city, but more recently many have come around to admitting that it makes sense to cluster our ever-burgeoning population in denser concentrations. As always, these visions reflected the preconceptions, and preoccupations, of the viewers.

At that moment, approaching the city by water and, as I had been more and more, thinking about how water threatened the shore, it wasn't any of the old dusty metaphors that I dredged up. I thought of our apartment on Wrightsville Beach. While ours was a small barrier island, not a chunk of glaciated bedrock like Manhattan, and was home to three thousand people, not eight million, in that moment the two were the same.

It occurred to me that what I was seeing was similar to the sight I had seen while standing on Masonboro Island looking back toward Wrightsville: the same tall empty bottles stacked up on the same late-night dinner table. Just taller bottles and a whole lot more of them. But still waiting to be swept aside by the angry drunk.

Sandy would not prove to be the true angry drunk, but just as with my hometown there was little doubt it would come one day. Hurricane experts rank New York, despite the relatively low odds of a major storm, as the country's second most dangerous major city, behind only the hurricane bull's-eye of Miami and just ahead of New Orleans. This is not based on history but on potential. What a major storm *could do* to New York.

After I got off the ferry, I hiked down the west side of Manhattan, along the river. The weather was cool and perfect. Cops on horseback trotted next to the bike path, while tree swallows shot out over the river, snatching insects. Not long after I passed the Chelsea piers, I took out my binoculars and watched a family of brant bob next to some old rotten docks. New Yorkers have their New York, but so do the rest of us, the town belonging to outsiders—not all of us rubes—as well as its denizens. My New York, like many people's, is a walking city and that day, keeping to my theme, I stuck to the water.

At Warren Street I finally cut across the island toward the Bridge Cafe, the city's oldest actively operated bar, where Orrin and I planned to meet. I was a little late but there was Orrin, standing on the street corner in front of the bar, smiling. He waved off my apology.

"I could stand on this corner and people watch for hours," he said.

His ankle was a little gimpy from a recent fall, he told me, so we slowly made our way back across the island. We stopped for large coffees, our shared addiction. The only difference was that I stopped after one cup while Orrin kept going. We also treated ourselves to blueberry muffins, and as we did Orrin talked about the new book on sea level rise he was finishing up.

"If the ice of Antarctica and Greenland continue to melt we'll see the most radically changing shoreline in thousands of years," Orrin said. "The latest IPCC report on sea level is very conservative, saying that we can expect a sea level rise of up to two feet over the next century. But it doesn't factor in what is going to happen in Antarctica and Greenland, and all indications are that the ice melt is increasing in both of these places. A more realistic assessment comes from the state of Rhode Island, a state that obviously has a lot invested in getting the estimate right. They're assuming that the rise is going to be between three and six feet."

He gobbled up what was left of his muffin and chased it with a shot of coffee.

"I think the rise will be greater," he said. "I think once the ice starts to melt things will go fast."

Properly fueled with coffee, we made our way over to the World Trade Center site. Orrin mentioned that "Ground Zero"—a name that he pointed out would take on a whole new meaning in a hurricane—was, like most of lower Manhattan, little more than five feet above sea level, which would soon enough put it, if you accepted Orrin's math, about two feet underwater. We climbed the stairs and stood on the overlook above the Trade Center site. Orrin seemed impressed by the size of the project, but I was actually a little disappointed for him. I thought back to the last time I'd visited, just a few months before, when the site looked like nothing so much as a vast archaeological dig, a hidden city being unearthed. I regretted that Orrin had not been there that day since he, as a geologist, would have appreciated the revealed bedrock and the thousands of unearthed cobblestones that had once scraped across and formed the island twenty thousand years ago. That day I'd been in awe of the sheer size of the hole in the heart of the city, with dozens of cranes and hundreds of orange-vested and hard-hatted men and women working far below the streets in a gap that must have been a couple of football fields deep. Down below a bridge spanned the gap, like something out of the Mines of Moria in Tolkien.

Maybe the 9/11 site should have been like church, where only solemn thoughts are allowed, but staring down at that hole and realizing just how close I was to the confluence of two rivers, and to the sea, and that where I stood was less than my own height above sea level, I pictured the gaping hole as the world's largest and deepest swimming pool. And if it was a pool then the skyscrapers surrounding it looked ready to dive in. All those years after the attack and the scene remained primal and chaotic.

The chasm had now been closed, but Orrin pointed toward the Hudson and described how its river waters would naturally seek out the low ground of downtown. As we toured New York City, he had me pay attention to the straight streets crossing the island, explaining that while they were good for finding your way in the city, they were bad for flooding since they ushered the water in, almost exactly the same mistake made on barrier islands, where streets were often built straight from ocean to sound. With Orrin as my guide, New York became a more elemental city, and he noted the dynamics of the Hudson, how if the tides and winds were right the water could back up over the wall at Battery Park. He told me that in Bangkok the stairs to subways were required to go down, then up again, before finally descending, so that stairways would not turn into waterfalls.

"The truth is that as sea levels rise it won't even take a hurricane to flood lower Manhattan," he said. "A strong enough nor'easter will do the trick."

For the rest of the afternoon we walked the city. I worried about his ankle.

"No it feels fine," he said. "I'm loving this."

It would occur to me that day, not for the first or last time, what a great traveling companion Orrin was. I have always been a father-seeker, but tend toward the wise and white-haired.

Orrin was more Gimli than Gandalf. He also, in his bluntness and direct humor, reminded me of my own father, who had died almost two decades before.

It was a day I would think back on often during the coming storms. Orrin pointed out how, just like on the Outer Banks, the cross streets would turn into inlets. He described how the river could become backed up from winds and tide. He opened my eyes to the vulnerability of the place: millions of people on a low-lying island. He described just what a big storm might do to such a place.

Three years later Hurricane Sandy more or less followed Orrin's instructions.

—

As Orrin, Jeremey, and I walked through the city after Sandy, we thought of another city. New Orleans and Katrina were very much on our minds.

"Think of the evacuations," Orrin said. "The refugees fleeing to Texas and to any other place where there was dry land."

He paused and pointed up at the skyscrapers.

"Imagine evacuating this place," Orrin said. "Evacuation is always the key, at least from a human point of view. Each place, and each storm, is different. Take Katrina. Katrina was Katrina for very specific reasons, for instance the direction it came in from and its duration. But in a general sense we are going to have a lot more Katrinas because we have a lot more development close to the shore. Think of the Florida Keys. Evacuation is close to impossible. The same is true in Tampa and St. Petersburg. A storm hit there in the 1800s that left parts of those cities under ten to twelve feet of water. Imagine that happening now with no way to evacuate millions of city dwellers."

During our trips on the Outer Banks, Orrin had pointed out the challenges of evacuating an island. It really came down to time: in the case of the island of Ocracoke, for instance, the ferries, the only means off the island, would have to shut down well before the storm itself hit. Residents, who had heard the weather forecasters cry wolf before, might choose to stay on the island, which, if the storm angled the wrong way, would prove deadly. This was an enormous problem on an island with a couple of thousand residents. What kind of problem would it be on an island of eight million?

"You almost don't want to think about it," Orrin said. "The bridges are so high and vulnerable here that you would have to shut them down well before the storm itself hit. Of course the ferries couldn't run once the water kicked up and the tunnels would be vulnerable to flooding. Imagine the traffic during a Friday rush hour and then imagine every single resident trying to get off this island."

Orrin pointed to the sky.

"If you really had sustained flooding, there would be only one way to evacuate. *Up.*"

Which at first doesn't sound so bad: there is plenty of up, plenty of high ground, in a city full of skyscrapers. But how long could you last camped out without power?

I imagined a stronger storm than Sandy, which after all wasn't even a hurricane when it hit New York. I pictured a fifteen-foot wall of seawater pouring through the Lincoln Tunnel. This is not just a Hollywood fantasy, but a real possibility. The truth is that to an experienced eye the whole island is almost preposterously vulnerable. In some places in this city, most pointedly lower Manhattan, the land rises only five feet above the sea. Put that together with the more conservative predicted sea level rise of two feet over the next century and you have a disaster in the making; put it together with the more bold predictions of a six- or seven-foot rise, based on the melting of the ice sheets, and lower Manhattan is already underwater. Then add in the fact that not long ago climate models at the Center for Ocean-Atmospheric Prediction Studies revealed that sea level rise in the northeastern United States will exceed mean sea level rise by about 8.3 inches, due to thermal expansion and the predicted slowing of ocean circulation in the North Atlantic. In other words, the sea will rise more dramatically here than in most places around the world.

—

At the end of our long day we drove Orrin out to a hotel near LaGuardia, which he pointed out was only ten feet above sea level. To give a sense of what this means he asked us to consider the fact that storm surges for a Category 3 hurricane can top twenty feet.

Orrin was flying home tomorrow, but after he got settled he joined me in my room for our ritual beer.

We talked about how we were encouraged by the fact that suddenly people were at least talking about ideas that no one had talked about during our last trip to New York. We compared this, hopefully, to the then-recent changes in attitude toward gay marriage. It was nice to see an issue reach a tipping point, and then actually tip. Maybe what was truly surprising wasn't the fervor to rebuild on the coast—that is same old, same old—but that this time there were actual murmurings about not rebuilding, and that three weeks before, the murmurings had become official, an offer by New York State, backed by the governor, to buy land and homes that were in the danger zone. True, no one was taking the governor up on his offer quite yet. But still. Maybe something had shifted. Even the vain Weather Channel jocks, so proud to always be the only ones standing out in the storms (and so outraged when mere civilians dare to go outside and stand next to them) were suddenly admitting that things like climate change and stronger storms might just be real after all. They had only recently been allowed to use those very words. (Sandy would mark a large change in the acceptance of climate language on the Weather Channel.)

Sandy, for all its uniqueness and duration, was hardly the strongest storm in recent years, nor the strongest we can expect. But as a wakeup call it had a few things going for it, not least the fact that one of its main targets was the nation's media epicenter. I found it encouraging that for the first time many people were acknowledging the price attached to where they had chosen to live: that it isn't the train's fault if you build on a train track. Things that were farfetched were now becoming accepted, or at least talked about. Sure, people were rebuilding—with energy and slogans and fervor and military metaphors abounding. But maybe the shift in consciousness, the sea change, had begun.

Maybe.

We decided that two beers were in order on our last night. If I had once seen Orrin as a journalistic subject, I had long since begun to think of him as a friend. We talked for a while more and then started to get a little emotional when we realized it was getting to be time to say goodbye. He told me that he and Sharlene were starting to look into some kind of group retirement home, which would mean leaving the house in Durham where they raised their five children.

"Boy, I'm going to miss our house," he said.

—

With Orrin gone, the trip was over—or so I thought. But the next morning Jeremy and I decided to drive out to the Rockaways anyway. There was someone there I wanted to talk to, a guy named Sal Lopizzo, who had spearheaded the local recovery after Sandy all but wiped out Rockaway Beach.

Quite by accident, we parked by the nursing home where the people had been trapped on the night of the storm, the one that was on the news. Waves had rocked the building, cut off its power, and flooded the first floor, terrifying residents.

Jeremy headed down the beach to take pictures. As I walked up the street from the nursing home, I ran into a man wearing a hoodie and walking a schnauzer. He was not reluctant to talk.

"FEMA was horrible," he said. "Their answer was 'go to a shelter,' but there was no public transportation. It was the apocalypse, like in a movie. Our police station was underwater. If you went up to 130th Street, a row of houses were on fire. Saltwater and sewage in all the houses. No power for weeks. We still can't get insurance since we went from zone B to zone A. And we saw sand in our nightmares."

He pointed to his place, a three-decker. We were not in Mantoloking anymore: here the buildings were crammed together and included many apartment buildings and, further east, high-rise projects. Both flames and water had ravaged the neighborhood on the night of the twenty-ninth, and after the man and I parted I walked along a block that had been completely wiped out in the fire, houses leveled with little evidence that they had ever been there except for the charred debris of human lives: scattered garbage and rusted-out bedsprings and cinderblocks. The water had stopped at the railroad tracks, but the only thing that stopped the fire was the social security administration building, which was made of brick.

It was on the other side of that building that I met Sal. Sal's black hair and New York accent were both thick, and at first glance, hunched in his leather jacket, he looked more hustler than do-gooder. But it was obvious he was whip-smart, and he knew everyone who passed us on the sidewalk as we talked, waving and saying hi and patting shoulders. He stood in front of the small YANA (You Are Not Alone) office, where he had just offered a class teaching locals how to get their contractor's licenses. Two weeks before Sandy struck, Sal had opened YANA, the idea being to help residents find jobs, and to retrain them for better jobs.

"The Rockaways had become a dumping ground. Where homeless people from prisons and mental institutions ended up. The block that burnt down was full of a lot of undocumented people. And a lot of the working-class people who are left feel stuck here. Like stuck in the muck. Everybody seemed like they hated each other. So what I wanted to do was to help them get jobs and better housing."

That had been Sal's original goal anyway, but when Sandy hit, YANA quickly became something quite different: the nerve center of a very local, very personal storm relief and recovery mission.

"We were cut off from the world," he told me. "We were our own little government."

He described how useless most of the big relief organizations were—"they handed out bottled water and potato chips"—and in contrast how great and resourceful Greenpeace was, parking their solar-paneled truck right in front of the store, providing the only power anyone in the neighborhood had for weeks. Then the Occupy Sandy volunteers moved in and became like family.

"We were feeding two thousand people a day. We made the building across the street into a medical center. We helped people get their medications. People from all over the city were bringing hot food every day. We ended up taking over four churches. We had makeshift barbecues, gave out clothes, diapers, all private donations. And small things, too: we were where everyone plugged in their cell phones thanks to Greenpeace."

An older Black man stopped by and asked Sal for three dollars.

"Don't chase women," the man said in a singsong voice. "Let the women chase you. You taught me that."

"That's right," said Sal, slipping him the bills.

Sal told me that the streets had been cleaned up only recently, just over the last week, in preparation for today's St. Patrick's parade. The holiday was two weeks off by my calendar, but he explained that the parade dates rotated through the city's boroughs, since there were only so many men in kilts to go around. And he warned me that I had better move my car before the parade started or I'd be trapped all afternoon. I did as I was told and drove the car a few blocks up the street.

It was on the walk back to Sal's place that I saw it. A sign.

There it was in huge letters painted on canvas that was being attached by two men to a fence in front of the Seaside Towers apartments. The sign read: ROCK JETTIES NOW!

The syntax confused me for a second but then I got it. It meant "Give the Rockaways rock jetties, and do it now!" I hurried toward the men who had just put the sign up and who were already climbing back into their double-parked van. They closed the door and I broke into a run. I caught them right as they were pulling out. I knocked on the side of their van. They motioned for me to come inside and I did.

I drove with them for a few blocks. John Cori was the driver, and the founder of Friends of Rockaway Beach, and he told me that his message was meant for Mayor Bloomberg, who would be walking in the parade.

"Look up this coast," he said. "All the other towns have rock jetties. They deflected the storm, slowed down the waves. Then look at our beach. We need protection, too."

After they dropped me off I walked down the beach and saw he was right. Feeble wooden groins (called jetties here), like rows of rotten teeth, were all that protected Rockaway Beach, while further down rock groins jutted out from wider beaches.

Police boats patrolled offshore and helicopters flew overhead. On the way back up the beach, I also noticed a guy hanging out in front of Seaside Towers. I introduced myself and he told me that his name was Aaron. He was living in the Towers now, but he had a house on 111th that had been destroyed in Sandy.

I told him about the sign I had seen out front.

"That's right," he said. "There's no sense rebuilding if you're not gonna protect the place."

Back along the parade route I stamped my feet to keep them warm and watched the kilted men march and play bagpipes and bang on drums. When Bloomberg walked through, there were some cheers, some boos. "Mayor Bloomberg: Make Rockaway's

Board Walk a Sea Wall," said another sign along the route. The message was clear. The rich may have gotten it worse down on the Jersey Shore, but here the poor were the ones getting screwed.

Who could blame these people for wanting to defend themselves? Orrin could, but I doubt he would have if he'd been there with us. *No humbling of reality to precept.* Walls may destroy beaches in the end, but in the short term they protect your home. How can you tell someone who lost their house when the ocean rushed up unobstructed that an obstruction is a bad idea?

Not their *second* houses either. While I was glad that people were starting to open their eyes to the dynamics of living by the sea, I didn't expect this change to proceed logically. It was natural for the residents of Rockaway Beach to want to wall off the ocean, just as it is natural for the leaders of New York City to start contemplating much larger walls, the defensive barriers at the Verrazzano-Narrows, Arthur Kill, and Throgs Neck.

One thing that Orrin had learned over the years: try to convince someone not to protect themselves and you are in for a hard argument. *Something there is that does not like a wall,* wrote Robert Frost. True, but something there is that likes it too. The hope of certainty, of protection and security in this world. The hope that you can wall off not just the troubling ocean but something a lot more frightening: the constant uncertainty of life on earth.

But walls breed false security even as they are doing their destructive and disruptive work. Meanwhile, hurricanes are defined by uncertainty: Who would have guessed, for instance, that the state Irene—until Sandy the only recent hurricane to hit the Northeast—would hit the hardest would be *Vermont*? You could go ahead and build giant barriers, but what if the next storm didn't adhere to the pattern of the last, coming from the other direction? Sue Halsey thought dunes were the answer on barrier islands but Orrin remained skeptical. What if the next time the winds swirled around from the backside?

It may be unfair that the Rockaways don't get the advantages of richer towns. But in the end hurricanes can be quite democratic in their randomness. And in the end we are all in the same uncertain boat. New Yorkers will now get to experience the same uneasy feeling that people in my state get every summer when the new hurricane season rolls around.

And yet we keep moving to the water even as seas rise and storms strengthen. Many things have changed during the millennia since human beings have chosen to live by the shore, but one thing hasn't: to live by the ocean is to know uncertainty. You can put up walls and try to block it off, but in the end the only realistic philosophy for those who choose to live near the sea is the one Peter espoused back at the Paris Café.

All you can do is take your shot.

PART IV.

A NEW WORLD

"It was like
A new knowledge of reality."

—Wallace Stevens

REALITY SHOW

I am standing on top of Bald Mountain in Sonoma Valley, staring profoundly off toward the far Pacific and the dying sun. I scrunch my eyebrows and squint a little, thinking this will add to my overall air of deep thoughtfulness. My facial muscles are responding not to the transporting magic of nature, however, but to another imperative: I need to look good for the camera. Cameras, actually. Off to my side, a young man named Jimmy, whose easygoing professionalism I've come to respect over the last ten days of shooting, points a surprisingly heavy (I have tried to lift it) camera at my head to film me in profile. Up above, a drone swoops over to capture the full grandeur of the moment. Following the director's instructions, I move closer to the edge of the mountain, striking a pose that is meant to suggest part world conqueror, part shaman. I am a New Age Cortez.

Later, the drone, in search of the perfect shot, will crash into an oak tree and be rendered inoperable. That will cheer me. Later still, when I see the drone footage from above on the TV show that it appears on, I will realize for the first time that I have a significant bald spot.

The idea of the show, put oversimply, is that nature is good and that screens—whether those of computer, camera, TV, or phone—are bad. But of course it takes a lot of screens to get this point across. The irony of this is the sort that is easy for a sixth-grader to comprehend. I know this for a fact, because when I use the TV show as a teaching moment with my then twelve-year-old daughter, railing against all her screen time and phoning and telling her she should get out in nature more, she will rightly point out that the whole time I was out in nature I was being filmed.

—

For over twenty-five years now, I have been a kind of reality show unto myself.

During those years I have frequently traveled to beautiful and endangered places to report on them. Almost all of these trips were taken so I could write about those places, and only one was to host a television show, but the reporting trips had something in common with the TV experience. Even without cameras pointed at me I was acting as both reporter and character, writing as I went, knowing that much of what I was doing would end up on the page. If I stopped to think about it, which I sometimes did, I would be aware of the sheer self-consciousness of the enterprise. Which was ironic, as I first went to nature to break free of the hall-of-mirrors that is my own mind.

When I was young, I romanticized *moments* above all else. Moments when I would escape my brain, moments when I felt free and, every once in a while, euphoric. I remember laughing when I came up with the epitaph I wanted carved on my gravestone: *He Had His Moments.* Another phrase, *the thing itself,* took hold of my young brain. This youthful sense that there was a reality below everyday reality was exacerbated by a high school reading of *Walden.* I still have my high school copy of that book, its spine held together by athletic tape, and can run my finger over the deep grooves under this sentence: "Let us settle ourselves, and work and wedge our feet downward through the mud and slush of opinion, and prejudice, and tradition, and delusion, and appearance, that alluvion which covers the globe, through Paris and London, through New York and Boston and Concord, through Church and State, through poetry and philosophy and religion, till we come to a hard bottom and rocks in place, which we can call reality..."

Reality. What was it? I wasn't exactly sure, but I knew it was different than the bullshit of grades and popularity and dating and sports and looming college applications that were the focus of too much of my young life. And I knew I felt closer to it while walking in the hundreds of acres of woods we were lucky enough to have behind our high school. My gruff businessman father loomed large in my life back then and I remember that one of those woods walks, which I took after blowing off my afternoon classes, culminated in a fit of wild happiness during which I tore a dollar bill in half, and where, for a moment, I was more Whitman than Thoreau. This purely adolescent moment came about courtesy of a purely adolescent brain, with its hippocampus not fully developed, which allowed for a kind of spiritual ecstasy that, drugs notwithstanding, I have not been able to recreate as an adult. And that sort of moment, like Wordsworth's "intimations of immortality," faded fairly quickly so that by the end of my college years I was already looking back and bemoaning my inability to retrieve that state. You could say that all of my trips since have been an attempt to return to that moment.

As pure as that moment felt, it was also bookish. Steeped in the Romantics and Thoreau, I believed such moments were possible, and so they were. It was not dissimilar to what other young people, influenced by another, older, book, might experience as religious. And it was religious, or at least spiritual, in ways I am not so sure my moments since have been.

Art soon took over, and the goal went beyond merely experiencing those moments and became recreating them. Soon the question was not just how I felt while staring up at the trembling leaf of a post oak or down at the shadow play of branches on a rushing river, but how to put those experiences on the page so that people reading them could feel as if they were staring up at the trembling leaf of a post oak or down at the shadow play of branches on a rushing river. This, arguably, was another layer between reality and me. I couldn't tell if my writing detracted from or enhanced my humanity. Thoreau knew the feeling: "My life has been the poem I would have writ / But I could not both live and utter it." This statement, sometimes taken wrongly as meaning that life trumps art, is bolstered by both the millions of words Henry spent recording his life in his journal and the fact that the line itself is written down. Talk about a reality show. As it turns out, cameras and phones aren't the only things that can distance us from nature. Our stories can do it too.

—

Part of my argument here, and my self-argument, is that it is hard to fight for a world when you don't *know* it, and love it, on a visceral level. One of my heroes, Doug Peacock, values the experience over the telling of the experience. I wish I was able to do so with more consistency.

We are quicker than ever to put screens between ourselves and the world. We narrate as we go, wasting little time to experience before we chronicle to the extent that the two become one. If what is wild is unplanned, spontaneous, and self-willed, then we have never been tamer.

Over the past couple of decades I have become a kind of professional moment-seeker. I have *uses* for my moments now, and almost as soon as they are lived I jam them into a book, just as I have jammed them into this one. In this sense my trips have been successful. All my running around has added up to a dozen books and a couple hundred essays. This is the worldly result. But in another sense they have been less successful: I can imagine my younger self regarding the older me as a failure, or at least a sellout. My moments are now the bricks with which I build stories, but at times I feel I have lost the moments themselves in the process. Or at the very least they are less authentic. Is this overthinking things? It sure is. That is the point.

It is not just the fossil fuel industries and their machines that are the enemy of wildness, or even those machines that we use to film ourselves or to stare at and type our furious and desperate communications into at all hours. Another machine that is certainly a culprit in the destruction of wildness is the nagging, distractible, fearful, order-craving, ever-hungry human brain. If living wildly is a discipline, which I believe it is, it is one that we, most of us, have failed miserably at.

I found myself first plowing this furrow of thought while spending a week in a landscape that many would consider, at first glance at least, a wilderness: the H. J. Andrews Experimental Forest in Oregon. Here was the very place where the spotted owl drama played out, a sanctuary for Douglas firs that were already growing when Columbus landed and that still stand like sentinels, hundreds of feet tall, over a moss-filled, rain-shadowed landscape, a landscape split by ice-cold creeks and topped by snowy mountain peaks. If this doesn't stimulate your brain's nature-romance center, then consider that I lived alone in a cabin and saw practically no one for most of my time there.

And yet it wasn't quite what you'd imagine. H. J. Andrews is a research forest, not a park, and research makes its own demands on the land, so that walking deep in the woods you would find a fallen fir with an aluminum ID badge on it, like a toe tag on a corpse, or, following a mossy path you might look up and see that you were being followed by a camera, as if you'd wandered into a slow-paced version of *The Hunger Games*. And the woods, though deep and beautiful and filled with the haunting drumming of grouse—a noise like a throbbing heartbeat—were also filled with what my scientist host called "research trash," the detritus of hundreds of experiments. "Research is an intrusive activity," he told me. It wasn't really trash, and it did nothing or little to detract from my nightly experience of having a beer by the creek while watching little gray birds—dippers—do deep-knee bends in the water, but it was yet another indication that we, *Homo sapiens*, were very much there, and that this place, however seemingly wild, had a *purpose*. Even the ospreys, familiar birds for me in an unfamiliar place, were involved in a reality show of their own, a TV camera pointing down into their nest atop a dead Doug fir to record their every preen and scratch. In the end, theirs turned out to be a particularly gruesome show, a little-too-wild version of wildness, when the most unmaternal mother cannibalistically devoured the corpse of an offspring that had starved to death. At that point more than one viewer, you would guess, turned off their TV, unable to stomach the spectacle of nature at its most raw.

For all this, it was a beautiful week, filled with socked-in sleeps of ten hours or more and the occasional sunny day when I climbed up Carpenter Mountain to stare over at the snow-topped mountains known as the Three Sisters, and the real problem of the week had nothing to do with the impingement of research on the world outdoors. No, the real problem was right there in my living quarters, and it was wholly invisible. My generous hosts had only recently managed to provide the place with an internet signal. This was kind of them, but by doing so they had invited the whole virtual world into my retreat. And so, each day, after my morning coffee, I would head not into the strange landscape outside my door, but into the familiar ether. I had a new book out, and my preoccupation that week was checking my numbers on Amazon to see how it was selling.

I was working within a tradition, and the tradition was not Thoreau's. The previous occupant of my cabin—or, more accurately, apartment—in the woods was a multimedia personality, composer of electronic hip-hop, and internet maestro named DJ Spooky. He had left right before I arrived, and the word at H. J. Andrews was that Spooky was a master of multitasking, constantly connected, usually to multiple online sources through phones, headphones, computers, and other gizmos, and that during his stay here, while staring up at the same moss-covered trees, he had at one point put in a call to Greece to rent out the Acropolis for the production of his new experimental play.

I am not saying this to be critical of the cabin's former inhabitant, but because I recognize myself in him.

—

If a tree falls and is not filmed, does it make a sound? My brief experience hosting a TV show about nature featured, not surprisingly, many staged moments. But during the filming, more than one authentic moment broke up through the grid of the artificial, like shoots of grass through concrete. Sitting in a canoe, for instance, preparing for my Kurtz-like journey down a river in Georgia to visit with the guy who has lived off-grid for twenty years, and who eats snake and otter for breakfast, I am left alone for a while by the cameramen as they prepare the drone for what I am told is the "beauty shot" of my paddling in. My job for half an hour is to sit in the canoe and wait for the crew to get ready, and while I do, swallows cut through the air above me and a patch of dark sky lightens to pink and blue. A pileated woodpecker flies in front of the boat, from one rotting tree to another, and it, so large from so close, looks almost like the ivory-bills that used to rule these woods. Lifted out of the particulars of the day, I forget I am wearing a mic and when the bird flies by again I let out a little hoot, which makes the sound technician laugh.

A week later, we are filming in the foothills of Salt Lake City, a long day that ends with my sitting in front of a tent with a beer ranting about Thoreau and Muir—all for the camera. But it is what comes after the filming that I remember. We are walking back to the car in the dark (the tent was a prop—I wasn't sleeping there that night) and I am helping carry the heavy equipment with the four other men, and we are walking quietly and the hills are dark with the city lights shining below and the stars shining above, and I feel a deep sense of mindless well-being. I am part of a small tribe instilled with purpose. I am a tired animal hiking through a beautiful place. I sleep well in my non-tent/hotel room that night.

Authentic moments pop up out of inauthentic frames. And so out of the unwild the wild can arise. We try to narrate and control, but always there is something in the world that works against it, that pulls the rug out from underneath our plans. How do we respond? Often by ironing the rug, then nailing it to the floor to make sure it can never be pulled out again. Rarely by laughing and acknowledging how boring life would be with only stable rugs.

—

Birds are the thing that most consistently helps me forget myself and remember the world. Birds are the thing I will miss the most. They are what most often surprise me, if not startle me, out of self.

Looking back, it is funny and not a little odd to admit that, in that thick forest of ferns and six-hundred-year-old trees that you could not see to the tops of, the most authentic nature experience I had was with an American dipper.

By all accounts, a dipper is a drab little bird. Gray, short-tailed, and sparrow-sized. Its song isn't much either, at least to my ear, a nice high whistling and trilling, true, but just okay for a songbird. Then there's its body, described variously by my bird books as "plump" and "chunky." Seeing it out of context, you might think: here's a bird that needs to start working out.

But that's the thing. You rarely see a dipper out of context. Context in the case of the H. J. Andrews forest meant a silver-flecked, fast-running western creek. And to see a dipper in that creek, or under it, is to see an animal supremely adapted to its surroundings. It is also to see a marvelous athlete, chunky and plump be damned. Because while American dippers may look drab, what they do for a living isn't. What they do is dive directly into turbulent, tree-strewn waters born of snowmelt. Then, both by swimming and walking on the creek floor, they feast on insects and insect larvae—on dragonflies, mayflies, mosquitoes, midges, fish eggs, worms, and even small fish—before reemerging and diving again.

If you are lucky, you might catch sight of a dipper readying for these dives. Their warm-up will appear to consist of old-time calisthenics, specifically deep knee bends, as they pop down and up again in the water. What they are really doing is looking for their prey, earning their name by dipping down and peering into the rushing creek, sometimes at the rate of once a second. They are aided in this by the nictating membrane that acts as an extra eyelid and allows them to see underwater and by scales that shut over the nostrils, as well as extra feathers, extra oxygen-carrying capacity, and a low metabolism that allows them to withstand the creek's cold. You could say that they are built to dip, or, looking at it another way, that the creek that they dip in is what created them, or their species, over the millennia. Either way, they perfectly fit their place.

Once it has spotted its prey, the bird dives. It is a moment of both commitment and transformation. The dipper's dark back turns silver as it submerges. It becomes part of the creek and takes on the creek's colors. What looked chunky above is sleek below. Some dives are longer while others are a series of rapid surface dives, and though I have not seen it, dippers are said to even sometimes dive from the air. As they go under, their wings still beat as if in flight but now through a denser element. If the dive succeeds, the dipper scrambles back to a rock or the shore with their dinner. If that dinner is immobile—an egg, for instance—the bird might place it on a rock and dive in again.

It is for the best that the dipper has ditched its old stodgy, English-sounding name, the water ouzel. It is a direct bird, after all, simple and pure of purpose. Running water is not just its dining room but its highway and you might see it hurtling up or down creek, wings flapping hard, barely a foot above the water's surface. Its domed baseball-shaped nest will be nearby, above the high water line, on a bank or below a bridge. But after the shared work of raising the young, dipper parents spend much time apart, and individuals carve out a "linear" territory that amounts to separate sections of each creek since the work of harvesting wild, running water is work best done alone.

If it so happens that a human being is watching the dipper's dive then another transformation may occur. Because in my experience the sight of *Cinclus mexicanus* in action tends to work a certain chemical magic on the brain of *Homo sapiens*. For one thing, any human watching the dipper dip is at that moment situated near a clear, relatively clean body of flowing water, since dippers are excellent indicator species and don't fare well in polluted creeks. And that creek will be somewhere west of the hundredth meridian on the North American continent. This alone may not guarantee a rising mood, but the gurgle, rush, and glug of the creek's music does make fretting harder. If the witness was born in the West, they may, with some justifiable pride, see the little bird as a kind of symbol, or flag, representing their region's waters. And if the witness is a stranger to the country, he or she may laugh with surprise when the dipper dips, regarding the little, not-so-drab bird as the miracle of adaptation that it is.

—

Dippers help me fight for this world.

Of course those who decide the fate of actual nature usually live far from it. It is the colonial way. And I know that what is decided in those unwild corridors of power is more important to the world than my communing with a small bird. Advocacy, and hopefully legislation, is imperative in our fight to slow the warming of the planet. The myth that it is individual action that counts most is one the fossil fuel companies have always nurtured. Individuals can drive electric cars and put up solar panels, but if governments don't take action, real action, all is for naught. I believe this and fight for it. But during my one stint on planet earth I have also tried to establish a relationship with the so-called natural world, whatever my own limits and that world's reduced state. This does not matter as much as passing effective laws that reduce the use of carbon-based fuels. But it still, I contend, matters.

Wild is not just acorns and bugs. Wild is the surprising, the unexpected, the funny. Wild is understanding just what a comically small role you play in the reality show called *Earth*. To be wild is to acknowledge that we are just another animal in a world full of animals, an idea that many claim to hold but few live by. To be wild is to at least attempt to think beyond the human, to think biocentrically not anthropocentrically, even while stuck as we are in the Anthropocene.

I have wandered pretty far afield, a long way from moments. So let me return us one last time to the H. J. Andrews Experimental Forest, home of research trash, dippers, Douglas firs that rise hundreds of feet high, and the former temporary home of both DJ Spooky and me. Before I walked through that landscape, I thought I knew moss. But I was just talking Vermont moss, Carolina moss, deep Pennsylvania moss. What I saw in the rain shadow in those Oregon woods was a thousand types of moss, moss that drooped and mottled and dripped. Brittle florescent moss and white wispy beard moss and shaggy green moss that looked like a tacky '70s carpet. And there, amid this mossy world, in the understory below the giant trees, after I picked up the rough-skinned newt (before I learned its skin was poisonous and began washing my hands over and over like Lady Macbeth), grew a small Yoda-like tree, covered with shag and yes, moss, that was called the Pacific yew. It is a useless tree, some would argue, certainly not the crowd-pleaser that the big firs are, but from that tree's bark, which is only a millimeter thick, was a substance found nowhere else on the planet, and from that substance scientists not long ago created Taxol, a drug used to treat ovarian cancer. I would argue that the tree has a purpose far beyond the human, but in the Anthropocene your chances of surviving are better if humans need you, and it turns out we need this once-useless tree. The yew has become a symbol of biodiversity, of the value of not destroying things even when we don't know they are valuable. Of the uses of the useless.

It goes without saying that a virtual yew tree does no one any good. You can't recreate it once it's gone. In our wild places things can happen that can never happen in a lab or boardroom or even, god help us, on a computer screen. The lab called the natural world has been conducting millions of experiments on a trial and error basis for billions of years. That is one selfish reason we need to protect our wild lands and parks and other preserved lands, and maybe connect them and make them larger. So that the experiments can keep on running. It makes no sense to burn down the lab.

What I am trying to say, I suppose, is that whether we can see it or not there is still a world beyond our screens and there might just still be something that world can offer us.

When I am done with this paragraph, I will check my email, then maybe head over to Facebook for a while. If you want to learn more about me, you can Google my name and find the TV show about nature I've been referring to. There are times when, caught up in my own plans and ambitions, I think I am pretty important. But I am not an idiot. I am not silly enough to mistake my own personal ecosystem with an infinitely greater one. That there is life beyond the personal, beyond the virtual, beyond the human, beyond *use*, is something that we can consciously try to ignore. But we ignore it at our own risk and to our own detriment. We cut ourselves off from life. We lose our connection to what evolves, to the still-wild world.

OCEANS AWAY

The dolphins cut back and forth in front of our bow, rolling and jumping and twisting. They spotted our boat and began to chase us as we chased false albacore, and before we knew it a half dozen of them had caught us and were weaving and darting in the slipstream of our bow. The three of us on the boat are laughing and smiling and pointing. The dolphins seem like sleek packages of exhilaration. Joy embodied.

It is November 2, 2021, and this is the first stop in a trip south to Florida that I am taking with my usual traveling companion, Mark Honerkamp. Hones, as I have forever called him, lives to fish and today he is in heaven. We launched from the dock in front of my friend Douglas's house. Douglas Cutting, a former student of mine and former fishing guide, is part owner of an old wooden house that stares out at the Intracoastal Waterway in McClellanville, South Carolina, a small fishing village about three hours south of Wilmington.

Before the day is over Hones and Douglas will have caught seatrout, red snapper, black sea bass, gag grouper, pinfish, and an oyster toadfish. The fish they don't catch, the false albacore, or albies, will provide the day's greatest excitement, rippling the surface with their sleek backs as they devour pods of baitfish, but refusing to take the bait we offer. Even I, a non-fisherman, get into the pursuit of the albies, standing up as the lookout in the crow's nest as I scan the waters with my binoculars. On top of all the fish, we are treated to the sight of bald eagles, a loggerhead turtle, and a vast variety of birds from scoters to little blue herons. But for me it is the sight of the dolphins bow riding that will remain the day's highlight.

When we first moved to Wilmington, I liked to tell people that I felt like I had moved onto the set of *Flipper*. One of the best things about moving south was suddenly having dolphins as neighbors. It turned out that my new home was not just the center for a large community of dolphins, but a hotbed of dolphin research. One of the most exciting discoveries about dolphin lives was recently made by three scientists, one of whom, Laela Sayigh, lived only a twenty-minute paddle from me when I first moved south. Laela coauthored a paper that concluded that bottlenose dolphins convey identity information with individually distinctive signature whistles. This may not seem particularly eye-opening until you stop and think that this is just a fancy scientific way of saying that they call each other *by name*.

You can try to deny the world beyond the human all you want. But face to face with dolphins it is harder.

—

I have somehow traveled far in this book without mentioning the large, wet elephant in the room. That is, I have mostly ignored the more than three-quarters of the globe that is covered by water. Thematically there is no excuse for this but stylistically there is. I have tried to make climate small and personal, and what is larger and less personal than an ocean?

But if the ocean's story is large and impersonal, it is also vital to our survival. And if we are overheating the world, this is where most of the heat is going. The heat and acidification of our ocean waters does not get the big headlines. Yet.

Dolphins, I suppose, are one way of making the watery and vague more specific.

If the oceans go, so do they.

After Douglas has docked and cleaned the boat, we stand around the kitchen island and drink cold beer while feasting on the fish we have caught and on stone crabs Douglas has prepared. We are joined by Greg, another of the house's co-owners, and his friend Drew, who has served as Greg's crew today in a king mackerel tournament. It is as if Hones has found his lost people. They talk fish, and I don't just mean they talk *about* fish. They are speaking a foreign language, one I understand only a little better than dolphin. Fishermen and hunters *know* things, and by keeping my mouth shut and listening I am learning some of those things, like, say, how to catch a stone crab through a kind of noodling. Douglas shows me the cut on his hand from the last time he tried.

"This is the best day of my life," Hones says at one point.

I laugh and tell him that the day my daughter was born was a little better for me, but Hones, as I say, lives to fish and so stands by his comment.

It is hard to imagine this sort of day may not exist in forty-two years. That it is hard does not make it untrue.

The next morning Hones heads out early with Douglas to fish for trout in the Intracoastal. When he gets back we thank our host and push off for points south. We drive through Charleston to Savannah, with me keeping half an eye on the clock. I have scheduled an interview in Tallahassee later this afternoon with Flip Froelich, a professor of oceanography at Florida State University.

I first met Flip at the dedication of Duke University's $6.75 million Orrin Pilkey Marine Sciences and Conservation Genetics Center. The building, in Beaufort, North Carolina, is part of Duke's Marine Lab. Though we were there to celebrate the building's opening, it sat close enough to the water to spur more than a few jokes about its fitting ending. What better finale for a building named after Orrin than going down in a storm?

During the ceremony, Orrin told the story of how he got seasick on his very first cruise on Duke's research vessel, the *Eastward*.

Afterward, at the reception, both of his families, blood and scientific, gathered around him. But his wife, Sharlene, was not there. Orrin had written to many of us a month before about her cancer and the news was not good. After sixty years together, there was now fear that there would not be a sixty-first.

I made it a point that day to talk to as many of Orrin's students as I could. The stories I heard were varied but the theme was the same: Orrin Pilkey changed my life.

"I remember sitting in his office, over in that basement in the old chemistry building," Flip said. "I was looking over at him through a tunnel of books and papers. I had been flunking out of Duke the year before and had gone to work for my dad, who was an auto dealer. A year in the shop convinced me I didn't want to be a mechanic. When I came back to school I was still lost. But Orrin talked me into coming down to the Marine Lab, and to going out on the *Eastward*. It was a life-changing event.

"Orrin's teaching style was unique," he continued. "I'd never seen anything like it. He would have us read everything about a subject and then he would teach by starting a discussion about what the articles all meant. Not just details. Big picture."

Froelich was now a nationally recognized expert on how changing climate was warming the ocean.

"There are probably two dozen stories like mine. People whose lives Orrin changed. People who are now scientists and coastal advocates."

After we visit Flip in Tallahassee we will drive down to Miami where I will interview Hal Wanless, a professor of geology at the University of Miami. Like Flip, Hal is a friend of Orrin's and has studied what will happen to the oceans in this era we are calling the Anthropocene. From what I have read, they offer up very different visions, but in the long run neither is particularly optimistic. The main difference in their pictures of the future isn't what is going to happen, but *when*.

—

It's a long way to Tallahassee and we are late. To make matters worse Flip and his wife are hosting a dinner party, though he graciously gives me an hour of his time at his dining room table. He does so with a kitchen timer between us, and when it goes off he has to check on the pot roast.

"Read all the stuff in the press about climate change and you'll see the oceans are still pretty much being ignored. But if we didn't have the ocean to take up the CO_2 and the heat, things would be about five times worse than they are. The whole big story is the ocean, and I get irritated when someone pops up and says we set a new temperature record in outer Saskatchewan. It's what's going on in the ocean that matters most."

I ask the obvious question: Why?

"We are naturally heading toward an interglacial period. No more ice in northern latitudes. The natural process of the last two really strong interglacials had Greenland melting down basically to a couple nubs. You're talking about three hundred to five hundred years until this happens again. We're probably accelerating it, pushing the curve faster than nature would be doing it. We don't know how much faster. The ocean absorbs at least two-thirds of the heat in the atmosphere, maybe closer to 90 percent. The ocean is going to soak up as much as it can and then it will stop. And that is what everybody in the oceanographic chemistry field is concerned with. Where is the threshold going to be in the upper ocean for CO_2 and acidification?"

Acidification, I know, means the death of coral reefs, a relatively lifeless ocean, the loss of plankton. I ask Flip my question about forty-two years, but he thinks in numbers bigger than that and in timespans longer than single lifetimes. He is no doomsayer, or rather he is no short-term doomsayer.

"Life will go on. The human species will adapt. Humans will move away from the coast, just as they have been doing for five thousand years in the Mediterranean. So I'm not one to get up on my soapbox and say it's the end of the world, it's a catastrophe. Though some parts of the planet, like the Bangladesh Delta, will become uninhabitable. It will be underwater."

I mention the Mississippi Delta and he agrees that southern Louisiana will be underwater as well. Then we talk hurricanes.

"I personally don't believe that the strength of landfalling storms has changed. The real change is in sea level."

I ask about the Antarctic.

"I believe the ice on Antarctica is safe for a couple thousand years."

"Well, that's good news."

"Now the peninsula, which sticks up much farther from the South Pole, is a different matter, and there will be environmental changes, but I believe the ice is safe."

The reason he believes this is worth pausing on. He explains that the continental shelf on all of the continents except Antarctica slopes down and it slopes down because as the ice came off, the continent rose up. The Antarctic is shaped the other way around, so these glaciers as they come out run into a high, not a low. People have argued that once the meltwater gets underneath and lifts them they will slide. He explains: "What the geophysicists are saying is, 'I'm sorry you don't understand the geophysics of the earth, as soon as you start unloading the ice, the edge lifts and the lift is going to be three times faster than what the melt is going to be.' This will slow melt. So I'm not worried about the Antarctic."

He pauses and admits: "Now you'll get exactly two opinions about that."

Flip belongs to the more cautious camp. Tomorrow in Miami I will hear the opposite opinion.

"The real culprit with rising seas is not melting glaciers, but heat. That is what accounts for at least two-thirds of sea level rise. The faster you heat the atmosphere the faster you heat the ocean and the faster sea level goes up. Only 10 to 20 percent of sea level rise is due to new fresh water from melting glaciers.

"The ocean is on about a twenty- to thirty-year flywheel. Heat that we put into the ocean twenty to thirty years ago is now expanding and driving sea level rise. Most oceanographers say that if you don't solve a problem now, you're going to pay for it in twenty to thirty years. But even if you solve it now, you'll pay for it in twenty to thirty years. There is that long a lag time. It is already too late for that.

"The natural cycle is that we're going to melt all the ice in the northern hemisphere like we did during the last two interglacial periods, whether it's in three hundred years, or five hundred years, or a thousand. It's all a blink geologically."

—

Down in Miami, Hal Wanless's climate future proceeds at a more brisk pace. He is not afraid to answer the question of how Hadley will fare when she is my age.

"I can imagine a four-and-a-half-foot sea level rise by then."

Where we live, that means many of the houses will have water lapping at their doorsteps. Lower Manhattan, where Hadley is moving next year for college, will be even more dangerous.

"*This* place will be underwater," Hal says, pointing downward.

This place is Bob's Burger's, a restaurant off of the Granada Golf Course in Coral Gables. It is five miles inland but only four feet above sea level. We take a seat in the corner.

"It already floods the golf course during king tides," he says.

King tides are extreme tides that occur when the sun, earth, and moon are in alignment and the moon is at its closest point to the earth.

Today is not a king tide, but the streets of Miami are flooded. Hones and I drove down here through near-blinding rain and almost obeyed the blinking warning sign that said: "Turn Around, Don't Drown." We didn't drown, but as we entered Miami we could barely see where we were going, and our car left a wake. The Organization for Economic Cooperation and Development lists Miami as number one among the world's ten largest port cities in degree of endangerment from flooding and sea level rise. Despite the warnings, Miami keeps building.

We begin by talking about diversions of the Mississippi back in Louisiana. Hal was an early proponent, after Katrina, of diversions and is happy that they are being built. Which leads to an obvious question:

"Is there anything comparably hopeful in Miami's future or is it a non-future?" I ask.

"It's a non-future," he says bluntly.

"Look, we have already had a foot of sea level rise since 1930. So groundwater is a foot higher. Then we get king tides on top of that; those king tides lift the water 3.6 or 3.7 feet. They can raise the water level four feet. And we're only four feet above sea level. And all this is *without* a hurricane."

Earlier Flip Froelich had said there were "exactly two different opinions" about Antarctic melt. Hal's is the other one.

"We are not looking at one to two feet of sea level rise by the end of the century. We are looking at six to ten feet. It could be more like fifteen to thirty feet. It is impossible it will be less than six.

"When we came out of the last ice age, the seas rose four hundred feet. It didn't just happen all at once but in a series of pulses and pauses. Every pause would build a reef or barrier island and every pulse would drown it. Some were ten-meter pulses, some one meter. That is how sea level works in the presence of ice in a warming world. What Jim Hansen calls 'ice disintegration.' When it goes it goes. And we all know that. We've all watched ice melt.

"What we've seen in the past is really our only guide for the future. And none of the models really incorporate that. The IPCC hides behind its models. There aren't many people that think beyond, that have what I would call a *geologic frame of mind*. Jim Hansen is one. The models are fine but they are lacking. Use your head and put feedback in that the models can't, and you get the reality. Consider that CO_2 is rising two hundred times faster than it did when we came out of the last ice age."

He is speaking quietly, calmly, but his message is urgent. It may seem that his words and message are the opposite of Flip's. In some ways they are, in some they aren't. Two visions. Not so different. It is only a matter of time. I think back to Chaco Canyon and our inability to think beyond our own small blink.

"People say, 'If we do something now we can turn this around.' The problem is that over 90 percent of our extra heat is transferred to the ocean and that is not turn-aroundable. Because you've got to first cool the atmosphere before you even think about cooling the ocean. Scientists say if we stopped burning fossil fuels today it would take twenty to thirty years to start cooling the ocean."

On this at least he is on the same page with Flip. But he takes it in a different direction. When I describe what Flip has told me about the geophysics of the Antarctic, he pulls out a napkin and draws his rebuttal.

"There is going to be rebound, but it is not going to be big enough or fast enough. That will show itself *this* decade.

"And of course we won't stop using CO_2, and it will keep transferring to the ocean. It will keep warming and warming for centuries. There is nothing turning around, even if we start today. We have already had enough warming for what will probably be a catastrophic melt of Greenland and Antarctica. That is unavoidable. The last time we were at four hundred parts per million in CO_2 was three to five million years ago, back in the Pleistocene, and at that time we were over twenty meters above the present level.

"With just a six-foot rise, only 44 percent of Miami-Dade County will be left," he continues. "And of what is left, three-quarters will be less than two feet above sea level. This place is ridiculously vulnerable. The sewage treatment plants are compromised. The roads constantly flood. All this without a hurricane.

It goes without saying that sea level rise will make storms expo-nentially more dangerous."

He holds up his hands. "And we're still growing like crazy down here."

We circle back to Hadley.

"So I would say that by the time your daughter is your age, most East Coast cities will be flooded. We are talking about an eight- to ten-foot sea level rise by the end of the century. Maybe eleven to thirteen."

I mention the fact that the world doesn't end at the end of the century.

"No, and we are just talking about the East Coast of the United States. And we are just talking about sea level rise. We haven't even talked about other aspects of climate change that are more important than sea level rise. Like high levels of methane and the acidification of the ocean. Things that are more important than 'We have to move out of Miami.'"

After I thank Hal, Hones and I drive west on the Everglades Parkway, also known as Alligator Alley. We pull over into the northern edge of the Everglades. Hones watches the alligators for an hour while I focus more on the wood storks roosting in the trees.

It is lush, abundant, full of life. Not just animal life, but rare and endangered plants like tropical orchids. We cross the bot-tom of the state and enter Everglades City.

I note that the elevation of the city is three feet above sea level.

BENEATH THE ICE

My mother is dying and the world is warming.

Both occupy my thoughts, but neither as much as you might think they would.

In fact, the two subjects share this: I prefer not to think about them.

Repression is my business.

And yet I can't close out the world entirely. The latest headlines say a large chunk of Antarctica is melting, favoring Hal's predictions over Flip's. I try not to think about this too much for the moment. But I read the article anyway.

The line that stays with me is the one that says this: there is movement *beneath the ice*.

—

On my very last day of fall term classes, November 30, 2021, I run away from North Carolina. I fly, happily, west.

I make the masked flight and land in Flagstaff, Arizona, then drive to my hotel room where I teach my last class of the semester. By Zoom, of course. I have resisted the technology as much as possible and taught almost all my classes this term in person, but today I am Virtual Dave, a fact that allows me to be back in a place I love, the American West, two thousand miles away from Room 1004 in Kenan Hall.

How do we respond to the encroaching darkness? Some of us throw ourselves into action, even if it doesn't help. Sure enough, the next ten days will be action-packed, and I have decided to unite them with the gluey cohesion of theme. My theme? I will follow the water. More specifically, I will, whenever I can, stay close to the Colorado River, the lifeblood of the American West and the river that provides close to forty million westerners with their drinking water. Not a particularly original theme perhaps, and one I am not entirely committed to. My friend Zak Podmore once paddled the whole river from source to sea in a kayak. I will not be doing that. My vehicle of choice will be a rental car, and, when possible, a small plane, and rather than following the river in its entirety I'll just be checking in with it. I am, as always, inconstant.

Still I think it is important to see how my old wet friend is doing. Water is on everyone's mind at the moment, and over the last two months this now decades-long drought has exhibited a new and terrifying symptom. This year is the latest recorded first snowfall ever on the Front Range in Colorado. As of today there has been no snow. None. Not an inch. Which is no small thing when most of the water you drink depends on snowmelt.

I had a stopover in Dallas this morning and the flight from there was a fine warm-up for my theme. The gridded squares of north Texas farmland became something else as we flew into southern New Mexico. The grid ended and the land became both wilder and drier as it started to dip and fold in crenulated patterns. We barely saw a tree all the way to Flagstaff. No people either, or roads, for what seemed like hundreds of miles. The very definition of the arid West. A parched, sere landscape of pale brown, almost white, with streaks of red and patches of paltry snow in the high ridges, though there is no way those patches of snow, or the few winding creeks, could ever quench the vast land's thirst.

—

Water, water, every where,
Nor any drop to drink.
Here is something else that happened this year:

An iceberg the size of Rhode Island broke off of Antarctica. It floated off into the Weddell Sea, where explorer Ernest Shackleton lost his ship, the *Endurance*. Interest in discovering that lost ship is high (they will find it next March), while that in the berg is low. Ho-hum.

The good news is that the iceberg won't really affect sea level much due to the old melting ice cube theory, but, according to *Scientific American*, the "ice shelves help to slow the flow of glaciers and ice streams into the sea; so indirectly, the loss of parts of an ice shelf eventually contributes to rising seas, according to the National Snow and Ice Data Center (NSIDC). The NSIDC also says that the continent of Antarctica, which is warming at a faster pace than the rest of the planet, holds enough frozen water to raise global sea levels by 200 feet (60 meters)."

I understand why people prefer to think about the *Endurance*. Despite the tragedy, it is a heroic human story. What could be less human than ice?

I read the article about the iceberg in the morning and have mostly forgotten it by the afternoon.

—

So much depends on that winding trickle down below us. If you really think about it, it's preposterous. And precarious. That all that has been built in the American West, not to mention the power to do the building, is based on the meandering strand of water snaking through the canyon down below. A modern desert civilization, one in deep denial about what it really is, dependent on something so simple and primal. Take that strand of blue away and watch what has been built flake and crumble like a desiccated leaf.

About an hour ago we took off from the tiny airport in Valle, Arizona. When I decided I wanted to follow the river I wrote emails to two people right away. One was Bruce Gordon, the pilot who owns EcoFlight, which specializes in flying environmental groups and journalists over threatened land in his small

Cessna 210 airplane. I have flown with Bruce twice before, most recently taking a lap around the West from Aspen north through Colorado to Wyoming and on to Montana and then back. When I emailed him, I wondered if he happened to be doing any flights along the Colorado and whether or not I could hitch a ride. My next email was to Amanda Podmore, a young environmentalist whom I met in Bluff, Utah, when she worked for a group called Friends of Cedar Mesa, and who now works for the Grand Canyon Trust as their Grand Canyon director. I asked Amanda if she, and her organization, would be interested in teaming up with me for a flight over the Colorado.

Amanda replied first. *Funny you should ask* was the gist of her reply. It just so happened that the Trust was sponsoring a flight over the Grand Canyon to the confluence of the Colorado and the Little Colorado on December 1, with EcoFlight no less, and they just happened to have an extra seat. Lucky me. I decided to immediately extract myself from the mire of academia and head west before the term ended. Meanwhile Bruce, up in the air no doubt, didn't reply.

Amanda and her colleague Amber drove us up to the Valle airport this morning. When we arrived we stood huddled and watching our own breath near the runway. Two journalists from Arizona newspapers joined us and waited for Bruce to land his little plane. And there it came, puttering out of the north, a speck growing bigger. Though I had not long ago spent three days with Bruce, most of it in the cockpit of his plane, and though we had dined in Lewiston, Montana, while bar patrons played bingo, and later gotten adjacent rooms at the Super 8, and though I had written about both that trip and my previous trip and sent Bruce both the books where I had written up those adventures, I was honestly not sure if he would remember me. That is because, while I like Bruce very much, he has a kind of gruff, distracted, no-nonsense style that left me unsure if my existence had really registered with him.

I was relieved to find out it had.

"Good morning, David," he said after climbing out of the plane.

He was not alone and he introduced me to Chris. Chris Benson was a younger man, in his thirties I guessed, and it would turn out he was a pilot, too, a pilot who was auditioning for a role flying for EcoFlight.

The three of us stood apart from the others for a moment, out on the tarmac, and I decided to make my play. I have learned, over years taking these sorts of trips, that it pays to be assertive, maybe even to the point of pushy. No, that's not right. Not pushy really. What it really pays to do is take a chance.

So I did. I asked if there was any possible way that, once the official scheduled morning flights were over, I could fly back with them to Page, where they were stopping before heading home to Aspen. That would give me a chance to look down at not just more of the river, but the reservoir, Lake Powell, which I knew had shriveled during the drought. As for how I would get back to Flagstaff, where my rental car would remain, I'd figure that out later.

"We're actually going to Kanab first to refuel," Bruce said. "Then Page."

"I'd love to fly to Kanab, too," I said.

We decided to continue the discussion after the morning flights, but at least the wheels were turning. I felt the old excitement. The possibility of spontaneous adventure.

The approach to the Grand Canyon by air was much like any approach to the Grand Canyon anyone has ever made, even, no doubt, the very first approachers. It seems almost as if the place were created to take you by surprise no matter how many times you've seen it. After the high piñon forests comes the desiccated yellow world where you swear nothing could grow and no life could live. This is the surface of another planet and it is not a hospitable planet. Then the first hint of earth cracked open. Next the great gaping opening itself and the blazing layers of orange, red, yellow, and, this morning, the thousands of pink creases bathed in blue shadows. And down there, below the terraces and deep in the crevasse, a shimmering reflection of the water that runs through it.

I misidentify the first water I see. The chalky blue river below is actually the Little Colorado, and the distinction of its color from its big brother's becomes obvious a moment later when we fly over the confluence. Where the canyon opens up below, the lighter blue meets the rich emerald of the Colorado itself.

I remember the strange blue color from my time camping in Havasupai, deep in the canyon. On my long hike down into that canyon I scraped my leg, which proceeded to get infected, perhaps from my swim below Havasupai Falls. My days down in Havasupai were not the romantic ones that Ed Abbey had experienced as he lingered and lounged there for weeks back in the sixties. Instead I lay feverish in my tent, uncertain of how I would ever climb back out. Though if you had to be trapped somewhere you could do worse. *Havasupai* means "people of the blue-green waters," which is indeed the color of Havasu Creek and the falls, thanks to magnesium, calcium, and high concentrations of bicarbonate from the springs that feed the creek. This is the same color I see now looking down at the Little Colorado, and as with Havasu these chemicals form tufa and travertine, which in turn give the water its unique color.

As the two rivers meld and begin their westward veer, the color of the larger river wins out. Though those rafting below might disagree, the water's journey is not a wild one, but a trip from the giant dam above to the giant dam below. On the other hand, this is one of the few stretches where the river gets to really breathe.

As for the question of who (and what) depends on the water below, without its flow you can kiss Las Vegas and Phoenix and LA goodbye. Good riddance, some rural westerners might say, but really? Societal breakdown wouldn't stop at those cities' borders. While the consequences of this river drying up are hard to imagine, it isn't just futurists who are pondering what water wars might look like fifty, a hundred, or a thousand years from now. And what about in forty-two?

From up here the Colorado still looks like a relatively thin trickle, and compared to true river giants like the Mississippi it is. And not just a thin trickle but a thin trickle that is abused, overused, and generally threatened at almost every mile of its long journey. Amanda and Amber have been narrating our journey through the large headphones that we wear like earmuffs and have outlined two of those threats. The first is the Pinyon Plain uranium mine, which we flew directly over on the way to the Canyon. There it sat, on Havasupai tribal homeland only seven miles south of Grand Canyon National Park, a small island of industry in a sea of pine. Although small, the mine is hugely dangerous since its bread and butter is an element with a half-life of 4.5 billion years that is poisonous to humans and that has already leaked into the groundwater. The US Geological Survey reported back in 2010 that fifteen springs and five wells were contaminated by uranium in the Grand Canyon region, and members of the Havasupai Tribe, who live inside the canyon, worry that Havasu Creek, their one source of drinking water, will be contaminated.

Worry is a too-gentle word, especially when we are talking about radioactive contamination spreading into the aquifer. Amanda informs us that the US Geological Survey has studied 206 springs and existing wells in the Grand Canyon since 1981—5 percent of which had uranium levels above the EPA's safe drinking water standards.

I am struck by how neat and modern the mine looks from the air, a near hexagon of fenced-in retention pools and outbuildings and an erector-set tower. This neatness is an illusion. While writing my last book about the West I snuck into the mine and found that what from the sky looks orderly, up close resembles a Rube Goldberg contraption held together with wire and strings and chewing gum. Back then I asked the same question I ask now: Who in their right mind would think it a good idea to build a uranium mine so close to the river that is the West's primary source of drinking water?

Amanda narrates the second threat as we fly over the Little Colorado. A Phoenix-based company, having had two earlier hydroelectric proposals on the Little Colorado rejected, is applying for a third for the Big Canyon, a tributary canyon adjacent to the Little Colorado River. Another genius idea. I'll let the Grand Canyon Trust's pamphlet, which they handed out this morning, take it from here:

> If approved, four dams will be constructed entirely on Navajo Nation land, against the Tribe's wishes. Billions of gallons of groundwater will be pumped from the same aquifer that feeds springs along the LCR [Lower Colrado River], depleting the river's water source during a megadrought. Additionally, the dams would lose more than 3 billion gallons of groundwater annually to evaporation—while one out of three individuals of the Navajo Nation are living without access to running water. The Big Canyon Project overlooks the concerns of Indigenous communities and will disrupt the spiritual and cultural practices of people who have called the Grand Canyon home since time immemorial. Sacred places where ceremonies are conducted, prayers are held, and people come to reflect and find peace would be destroyed by flooding, industrialization, and noise.

This in an era when the water is drying up and we are finally thinking more about taking dams down, not building them.

After we land and Bruce takes off with a second group, I head over to a local restaurant and eat a big breakfast of bacon and eggs over easy. I'm convinced the odds of me doing any more flying today are long, but back at the airfield, after the Cessna lands, I am surprised when Chris approaches me. Yes, they will give me a lift to Kanab, where Chris will pick up his smaller two-seat plane, and then on to Page. And what then? Then, Chris tells me, he will fly me from Page back to Flagstaff.

Really?

I have rolled a seven. It turns out that Chris once lived in Flagstaff and it is his home airport, where he learned to fly, and he would be happy to give me a lift back. What can I say? *Thank you.* Many times over. I also thank Amanda and Amber, then say my goodbyes to them before we take off and fly north and west.

The first surprise, after twenty minutes in the air, is the smoke.

"The whole world is fucked if we are seeing smoke in the Grand Canyon in December," Bruce says.

We are not talking a little smoke here. Great billowing clouds fill the southern canyon like a bowl, dipping and penetrating the great cracks.

As we fly north to Kanab, we leave one fire behind but soon spy another up ahead near Brian Head.

The sights are still spectacular, though, and there are the Vermilion Cliffs, with their swirls of red and white, and there is Navajo Mountain, huge and looming and as constant as a beacon, and there are the lonely Henry Mountains up ahead, and here in fact is the whole landscape of northern Arizona and southern Utah laid out before us like a map.

It is deeply absorbing to see the landscape this way, though to be honest I am slightly distracted from my gawking by what is going on in the seat in front of me. Chris, who has never flown this plane or one like it, is taking the helm, while Bruce coaches him. As comfortable as I have become in Bruce's plane, that is somewhat dependent on the fact that, until now, Bruce has always been the one flying it. When it turns out he will be letting Chris land in Kanab, I decide it is best to go back to focusing, as much as it is possible, on the landscape.

Chris of course sticks the landing, and once we are down I chip in for gas. Then Chris takes off in his much slower two-seater Cessna, heading to Page where Bruce and I will meet him. On the way to Page, we skim along the newly restored border of the Grand Staircase-Escalante National Monument, the same one that Trump reduced and Biden recently restored. Opponents of the monument like to claim that there is much "local opposition." Maybe, depending on how you define *local*. The western author and photographer Stephen Trimble told me: "Communities like Kanab bordering the monument have seen increases in population, jobs, personal income, and per capita income that mirror other western counties with national monuments or protected lands." Of course, the punchline of claiming that the locals oppose preserving land is in forgetting who the real locals are.

Members of the Utah legislature often ooze a patriarchal, patronizing attitude toward their Native constituents, and that was never more clear than when then–Utah senator Orrin Hatch, sitting by Trump's side as he signed the reduction of Grand Staircase and Bears Ears into law in Salt Lake City, said: "The Indians, they don't fully understand that a lot of the things that they currently take for granted on those lands, they won't be able to do if it's made clearly into a monument or a wilderness... Once you put a monument there, you do restrict a lot of things that could be done, and that includes use of the land...Just take my word for it."

Yuk.

In October 2021, Grand Staircase-Escalante National Monument, which contains some of the most remote land left in the contiguous United States, was restored to its original 1,880,461 acres, which makes it the country's largest monument. From up here we can see much if its southern edge: the small bump that gives Mollie's Nipple its name, the dry, deep streambeds of Paria Canyon, the Cockscomb, and the Grand Staircase rising like, well, a massive staircase.

And then we see *it* in the distance. That incongruous bright blue in the middle of the desert that signals Lake Powell. No matter how many times I have been here it is always a shock to see a vivid blue-green lake in a pale, sere world. Here is where the Colorado has been corralled, and where it pools and gathers before continuing its winding journey.

This time, however, there is a lot less of it to see. The first thing I note is a giant rock that used to be an island but that now sits on dry land. The next is the great bathtub ring where the reservoir has fallen, down from its red rim to its white. The rim is white calcium carbonate, but it really does look like what people have long called it. The tub is low and the rim is showing. Unusable boat ramps wallow far from the water. Perhaps Lake Mead, where water levels have dropped lower than they have since the Hoover Dam was built in the thirties, is equally dramatic, but I don't really know Mead. Powell is my touchstone, my measuring stick, my lead line. And for me this place makes one thing clear like no place else: it is low tide in the West, and high tide is not coming back anytime soon.

Bruce lands in Page and I climb out onto the tarmac. A runway attendant offers me a ride over to the pilot's lounge, where I am supposed to meet Chris, who is still puttering toward us at half the speed we flew. In the lounge I see a sign on a door that reads: PLEASE DO NOT DISTURB. PILOTS ARE RESTING. I gently open the door, find the room empty, and camp out in one of the reclining Lazy Boys. I turn the lights out and my lights soon follow.

I'm not sure how much time passes until Chris finds me,

but before long we are back up in the air. Chris's plane makes Bruce's seem roomy, and my knees have barely an inch of clearance above the yoke, or control wheel, which concerns me when Chris tells me to make sure not to hit it while we are taking off. But all is well and, after a brief detour to look down at the Glen Canyon Dam, we are heading south along the river. Chris is a great teacher but I am not his equal as a student. With a master's degree in geology, he spouts many fascinating facts about the canyon's evolution, none of which I will remember. What I will remember is the twisting river, a shining green line deep in the canyon, and the Vermilion Cliffs on the far side, and the sheer joyous majesty of the improbable landscape. I feel immensely lucky to be seeing so much of the river I have come to follow. It is an act of generosity on this stranger's part, to fly a few hundred miles off course to drop off someone he barely knows.

Soon enough we are seeing snow on the tops of peaks, and, rounding the mountains, landing in Flagstaff. Amanda picks us up and drives me to my rental car, and I decide to spend the night at the Hotel Monte Vista, where I am given, for reasons unknown, the Jane Russell room. The hotel bar is called the Rendezvous, and when I offer to buy Chris a beer he decides to call a friend and spend the night in Flagstaff. He'll fly back home tomorrow. More beers (Tower Station IPAs, to be specific) follow and we are joined by two writers and rafters, Zak Podmore and Kevin Fedarko, whose book *The Emerald Mile* is one of the best to come out of the West in recent years. Amanda will join us soon. We talk about our fathers' deaths, we talk about how Zak could write a great river book, then after the third beer I'm not sure what we talk about.

It is here water comes back into our story. Or lack thereof. The day has been so exciting that I have not drunk any. Not a single glass. A rookie mistake if there ever was one. At seven thousand feet, and after just over twenty-four hours in the West, I follow my beers, not with a clear glass of that lifesaving liquid, but with whatever that margarita-like drink is that I see Kevin drinking.

That is all I remember of my full day. I stagger up the stairs to the Jane Russell room, chug a glass of water at last, and sleep eleven hours.

—

In 2019 the Yale Program on Climate Change Communication and the George Mason University Center for Climate Change Communication conducted a national survey and consolidated the results in a report called "Climate Change in the American Mind."

One of their conclusions was: "About six in ten Americans (63%) say they 'rarely' or 'never' discuss global warming with family and friends."

Is this really a surprise?

Two of my closest friends often respond to my book and movie suggestions with a question: "Is it depressing?"

I don't blame them. We bury. We hide. We learn to ignore. To take deep breaths and buy coffee cups that tell us not to panic. We don't go down in the basement.

For the most part it works. For the most part we couldn't function otherwise. We focus on our jobs, our stocks, our portfolio. Only the maladapted continue to focus on the darkness.

—

I feel surprisingly good this morning. Ah, to be a strong, young, sixty-year-old. The flower of aged youth.

After an invigorating and cold stroll through downtown Flagstaff, I decide it is time for a more up close and personal encounter with the river. I email Zak and Amanda, apologize for any slurring or other sloppiness near the end last night, and ask them to recommend a hike.

After two cups of strong coffee, I am following Zak's directions, driving north and basically traversing the same course that I flew yesterday with Chris, though this time on the ground and in reverse. I look up occasionally to see if I can catch sight of his Cessna heading home. I wonder if his headache is as bad as mine.

I've always liked the drive up to Page, in the shadow of the great stone ridge to my right. At Bitter Springs I veer off west toward the river. This is the first place I ever saw a California condor soar, and today it will become the second as well. The Navajo Bridge over the river at Marble Canyon provides a magnificent view of the emerald river, a much closer look than yesterday. But at five hundred feet above the water it is still not close enough. Today I am determined to feel that water on my skin. To perform a kind of ceremony of reunion.

My first attempt to get close to the water is a failed one. I park along the road so that I can hike Cathedral Wash, which leads in great stone steps down to the Colorado. The sign at the trailhead makes it sound like a fairly standard hike, though there are hints of the challenges that will prove my downfall. Usually these signs, if anything, are like hysterics, overplaying the danger, but this one is a western stoic, underplaying. "Canyon walls rise as you follow mostly dry wash to the Colorado River. This adventure has no *major* technical challenges but *scrambling and careful decision making are needed to choose safe routes* over ledges and drop offs...*some climbing and scrambling required*" (italics mine).

It is three miles to the river, and at first I am not just happy with my decision to hike the wash but thrilled. Scrambling down the rock ledges, I avoid a muddy pool with a thin layer of ice. I'm heading east but my true direction is down. It isn't hard to imagine water rushing over these sandstone shelves, feeding the Colorado. The surrounding walls start at about my height but within a half mile they are thirty feet high and growing. I drop down into a magical world of red sandstone as the walls rise on either side of me, and at mile one I look up at what I first assume is just a vulture. But, no, it is not a V wobbling in flight, and sure enough when I lift my binoculars I see the massive wingspan and know I have seen my second condor. (My bird book tells me that the birds are so large that some people, seeing them for the first time, think they have spotted an airplane.)

A mile and a half later I wish I could fly, too, since that seems to me the only way, not mere scrambling or careful decision making, to descend the final great drop-off. I try the north side first, then the south, inching along the ledge and not looking down at the fifty-foot drop. I suddenly don't feel so young and strong, and when I try to pull myself up onto a ledge above my head, I fall backwards. When I stand up, I see blood splattered on the rock at my feet and, after inspecting myself, discover the cut on my elbow. Nothing too bad…and yet.

I'm not proud of my decision to turn around but I'm pretty sure it is the right one. This is not the sort of epiphany I was seeking when I started out this morning but it is the one I get. *You are older than you think.* And possibly more fragile. You have a kid and a wife. It is better, for the time being, to be alive. So, tail between my legs, I retreat.

Luckily, just a mile down the road I find an easier and happier way to encounter the river. Paria Beach, where the Colorado rushes below canyon walls, is the perfect spot for my baptismal dip. The water here really is emerald, the choppy white of the rapids a spectacular contrast with the river's rich blue-green. It looks even richer and darker in the shade of the rock wall, and lighter but equally spectacular in the sunlight. I have the river to myself it seems, right below Lees Ferry and the confluence with the Paria River, though today the Paria, brown and weak, does not seem to be bringing much to the party.

Downstream from me float about a hundred of what I at first think are ducks, white-billed and bobbing with butts in the air, feasting on something. My bird book informs me that they are in fact not ducks but coots, and what they are after is "aquatic vegetation." Makes sense. Enough birdwatching. Time to join them.

I wade out into the green rushing river, looking across the water at a great orange buttressed fortress of rock. It was nice to see the river from above, but better to be immersed in it. I have always had the rule of three for any body of water. My swims don't count if I don't fully immerse myself three times. Yes, it is fucking cold. But I feel wide awake and, despite my earlier cowardice, fully alive.

The next morning at the crack of dawn I leave my hotel room in Monticello, Utah, and head for the tiny Moab airport. Sometimes it pays to have that extra beer. Headaches are no fun but friendships can grow out of nights like yesterday's. Chris texts me the code for the gate and soon I am meeting him out on the tarmac. Thanks to Bruce I have already seen the diminished Lake Powell, but today, with Chris as my guide and aerial chauffeur, I will get a much closer look at the changes not just to the reservoir but to the Colorado itself.

Sightseeing is such an inert word. But it's sights that we see today, and some of these are of the legendary variety. The Confluence. The Dollhouse. The Loop. Island in the Sky. The Maze. Dark Canyon. All impressive, dazzling even, rock twisting into every imaginable shape in a land without people. But today's star is not rock but the disappearing reservoir and returning river.

"Lake Powell is the poster child for aridity," I hear Chris saying in my headset. "It is the melting glacier."

Today Lake Powell is operating at 26 percent of capacity, 168 feet below its full elevation of 3,700 feet above sea level. When the reservoir was full, the water backed up so that the river became a lake. What I didn't know until today is how far the "lake" extended up the river, and what its withdrawal has left behind as it dries up, which is basically a world of mud and sediment.

We fly over the Hite Marina, which is so far away from the water it no longer merits the name. A boat ramp to nowhere. Down below, where the lake once was, flows a slow, stagnant "pseudo river," in Chris's words, winding through mud banks and mud canyons. Chris estimates that there are now thirty to forty miles of pseudo river where lake used to be.

We fly above a delta of mud. What we see below looks more like a dry map image of a delta than an actual delta. A corridor of sludge.

From a river runner's perspective the changes are not entirely depressing, since the river is returning as the lake withdraws. New rapids, in places they have not been seen since the 1960s when the Glen Canyon Dam was built, are emerging out of the mud. One day soon this might be glorious, but at the moment it's a mess. Rafting guides are not just dealing with a new river but with a paucity of places to camp, since spending the night on a ledge of sludge is less than romantic. "It smells like dead fish," Chris adds.

The Park Service doesn't know what to do. They would act, maybe build new boat ramps, but they are not dealing with a stable situation and the river could quickly morph into something else, something unexpected.

"Every month it looks different," Chris says.

Chris tells me that his friend Mike DeHoff of the Returning Rapids Project calls the mud and lake sediment that we are looking down at "the tailings of the mining of the Colorado River for water."

That gets at it pretty well.

Lake Powell was supposed to last forever. It is beginning to look like forever is about sixty years.

"Hard to imagine this is the drinking water for nearly forty million people," Chris says.

—

Consider this:

There are more than four hundred lakes below the Antarctic ice.

Lakes! How can there be water below ice? Why doesn't it

freeze? Part of the answer is pressure: all the weight of the ice allows the water below to remain liquid below the normal freezing point. At the same time, the ice actually insulates the water.

Much hinges on this water that resides a mile or two below the ice cap and above the rocky bed of the continent. It lubricates, it determines how the mass above will move and flow. And how it flows could mean everything.

Didn't I say it was a beautiful world?

On the flight back north I get to see a sight I have only seen once before, and never from the air. The confluence of the Green and the Colorado is the heart of Canyonlands, and it doesn't disappoint. We cross near where I used to camp in Needles and then over the alternating red and white sand doodles called the Dollhouse. Down in the deep gouged canyons the two rivers conjoin, a beautiful marriage between the West's lifeblood and its chief tributary. The Colorado is a lighter green than the Green, which has its source in the Wind River Mountains of Wyoming, and which travels 730 miles to keep this meeting.

Not long ago this was also where the Colorado officially "began," since the river above this point was called the Grand, not the Colorado. In other words there was no river called Colorado flowing through Colorado. That changed on July 25, 1921, when Congress passed a joint resolution changing the river's name to the present one.

After we fly over the confluence we begin the trip back to the Moab airport. As for fuel, for those keeping track of my hypocrisy, this tiny plane burns about five gallons an hour, and we have been up here for two. Chris wrestles with this and dreams of electric planes, but if you are going to fly you can't do much better than his plane, gliders excepted.

We are nearing Moab with the La Sals on our right, to the east, when it occurs to me that I should check in on some old friends. Chris agrees to fly toward the mountains and Pack Creek, the home of Ken Sleight and the place where I witnessed the flash flood last summer. Unlike the Front Range, there is some snowpack in the La Sals, though Chris, who studies avalanches, tells me it is not just light but unstable, leftover from October and November storms. From up here you can see how the fire swept down and then up, burning all the way to Geyser Pass. The plane shakes with mountain turbulence but since Chris is nonplused, I decide to be too. A huge swath of dead scrub oaks now stands like a forest of burnt matches covering the sere mountainside. Chris points below at a yurt where his friend lives and tells me that he helped the friend evacuate as the fire approached. I stare down at the pink-roofed buildings and fields that make up Pack Creek and at the spot where Ken's incinerated Quonset hut stood.

Chris lands in Moab and we shake hands and say goodbye out on the tarmac. This, we decide, will not be our last adventure. It is only eleven-thirty in the morning, but I have a date on the other side of the divide back in Boulder. For those keeping score, my small rental car will use about as much gas as Chris's plane. Since I am in a hurry I will only stop once after getting gas. That will be near Fruita for a frozen plunge into the Colorado.

I am aware of my own hypocrisy as I drive. And I am aware that that awareness does not help. I am trying to remain clearheaded as I contemplate what is coming at us. But more and more I find myself conflating things that perhaps should not be conflated. Losing nature and getting older. The reduced number of birds and reduced expectations in my life. My mother, in assisted living, who no longer recognizes me. I shake my head to clear out darker thoughts. Plunging in cold water helps.

—

Time to get to the source.

During my years in Colorado I had two basic groups of friends. The first were writers at the University of Colorado in Boulder. The second were ultimate Frisbee players in the same town. Only one other person I knew overlapped both groups.

The Mark Karger that I got to know when he was still in his twenties was a wildman poet who worked in a tree-cutting service and who was not afraid of doing any drugs that came his way. He was also, unlike me, a born westerner who had climbed a dozen fourteeners and whose father was an outdoorsman who took his kids fishing up in Rocky Mountain National Park in the Grand Ditch near the source of the Colorado.

Today Mark, now a father of five, is pointing his truck up from his home outside of Denver and back toward the place he fished as a kid. Our goal is to hike in as close as we can to the source of the Colorado, something that would be impossible, or at least require snowshoes, in a normal year but that we think we can pull off in just our boots today. To get to the source during the so-called warmer months we would drive north and head through Rocky Mountain National Park, but that road is closed for the winter despite the lack of snow. (Maybe that will change if we keep experiencing winters like this.) So instead we drive up the back way (which is the front way for the river) through Winter Park, Tabernash, and Granby. We reach Grand Lake, which, like Grand Junction, Grand County, and a dozen other Grands, is named after what the Colorado was once called, and which, like every other body of western water I've encountered, is low, low, low.

Today's elemental goal is water, not fire, but what the latter has wrought can't be avoided. Not here. The Troublesome Fire, its understated name taken from Troublesome Creek, a tributary of the Colorado, laid much of the nearby landscape to waste. "Mountains' worth of forest torched," is how Mark puts it. InciWeb, an interagency website for the Fire Service, Fish and Wildlife, and almost every other land-oriented government organization, puts it this way: "The fire was fueled by wide-spread drought, numerous dead and down beetle-killed trees, red flag weather conditions created by high winds and dry conditions, and poor humidity recovery overnight. The combination of these factors led to unprecedented, wind-driven, active fire behavior with rapid spread during the overnight hours." Among other things, the Troublesome performed the impressive feat of leaping the Continental Divide from the west and charging down the east side.

It turns out Grand County is as good a place as any to write about the end of the world. Miles and miles of spindly charred trees mark the hills. Not just a burnt landscape, but one that in places has been incinerated to the point of sterilization. This was not a rejuvenating fire. Nothing grows in the fine duff amid the charred trees a little more than a year after the fire tore through here. Our feet leave prints and send up puffs of dust. Wind blasts through the dead forest and kicks up dust devils that swirl across the charcoaled field.

At the park entrance we talk with the ranger, Russ Smith. He has been here since 2008, and says the place has "become part of me." He was also here for the fire and its aftermath and has his own horror stories.

"What you're seeing now is bad, but earlier in the season was worse. We were having dust-nadoes that were going two or three minutes long and spouted two to three hundred feet high. At first you would look at it and think it was a column of smoke, and we all had PTSD, and we were like, 'Not again.' But it was actually a tornado of dust and ash. I've never seen anything like that grow that tall and last that long.

"The thing about this fire was the speed. I took the stuff in the house that I couldn't replace. I have several Ray Lewis jerseys. You start to thinking what's going to burn up and what is not. There was a spot fire across the street one evening. I was about half lit but that sobered me up fast. An officer banged on the door and yelled, 'Russ, we're leaving,' and I left with him. Right away. And after we left we didn't come back. We were gone for almost a month. We were exiled. And we really didn't know anything that whole time.

"It felt like the end of the world. The one good thing that got me through was that last spring the color came back. All that grass that you're seeing back there that is gold was green. Not in the super-hot spots of course, they were still burnt. But in a lot of places. And that's when the animals came back. Everything but chipmunks and ground squirrels. There's not much for them anymore."

We will not be hiking to the actual source today because the fire has closed the road from this direction and to get to the source would necessitate a twenty-six-mile hike. Instead we will make the six- to seven-mile round trip to the meadow that the drainage descends to, right below the source. Mark asks Russ if there is an actual body of water at the source.

"Not really," Russ says. "On most maps you see a small pool, beyond Skeleton Gulch. But I've gone up there in late spring and early fall and I've never seen it. It's more a swamp from the drainages than any kind of lake."

He points up at a line slicing across the mountains that looks like a road. "That's the Grand Ditch."

The Ditch, fourteen miles long, is how water gets to people on the east side of the Divide. It's an aqueduct built in 1890 that provides water to the Front Range, including my old home, Boulder. I have found myself railing against the damming and redirecting of the Colorado, the damming at Powell and Mead, and the aqueducts and canals that send it toward Los Angeles and Phoenix, but the Grand Ditch seems different, less offensive, even quaint. But is it really so different? Doesn't the overengineering of this river start right at its very source?

Yes, so why doesn't it bother me as much? Why doesn't it assault my sensibilities the way massive modern projects do?

Maybe it's just a question of scale. The Grand Ditch is hardly natural, but it does have more of a feel of a natural redirection, in part because it is driven by gravity. It seems like something Ryan Lambert could get his head around, though in Louisiana, unlike here, they are simply asking the river to do something it had done for thousands of years, before human jiggering.

After we say goodbye to Russ we follow the road up into the park until right before it is closed off, where we pull into the trailhead for the Colorado River Trail. Before leaving Mark's truck he opens the console and reveals a bag of mushrooms. My mushroom days are long gone, or so I thought until now. Given the circumstances, a small nibble on a stem or two might enhance the day's quest.

This December day at nine thousand feet is warm enough at first that we don't wear gloves or winter hats. We'd talked about snowshoeing when we planned the trip but there is no need. A thin layer of snow partly covers the path. The river, which we come upon fairly quickly, is iced over in places. We stop at a small bridge and watch it flow elegantly both over and under a thin layer of ice, listening to its fine burbling music. The mighty Colorado is about twenty feet wide at this point. It is hard to imagine the same stream quenching the thirst of Los Angeles.

The hike in, just under four miles, is beautiful but uneventful. We see an American three-toed woodpecker, chickadees, the prints of moose and elk. We have gotten a later start than we hoped and know we won't be able to linger too long at the meadow below the source. The trail grows steeper and icier as we close in, and I worry when Mark slips a couple of times. He had hip surgery in the late summer and this is the longest hike he has taken since. But as it turns out, I am the one who goes down. A classic pratfall on the icy downhill where my feet slide out from under me. I land hard on my back while my phone, which I've been holding to take a picture, goes flying six feet up in the air. Sadly, I am worried more about the phone, the machine on which I am recording my travels, than my posterior. But somehow both I and it go unharmed. The phone lands on the ice above my head and slides down the hill to my hand as if trained.

The meadow where we reencounter the river was once a

mining community called Lulu City, built in 1880 after silver was discovered nearby. By 1881 there were forty cabins here, but four years later, after concluding that the silver was low grade, they had all been abandoned.

The snow is deeper here, and the sun glistens off it. On the side of the mountain above us is the drainage that feeds the river. The Colorado flows down the mountainside and through these fields, mostly below a layer of ice, but then emerges to slide over frozen rocks and crusty snow. The sheen of ice reflects diamond patterns in the sunlight as the water runs and gurgles and races below, burbling back up. That most beautiful of all music, cold pure water over rocks, sounds even better this high up. "I love all things that flow," said James Joyce. Amen. At one point the Colorado is so narrow I can straddle it.

Mark warns me that we can't stay long. It may be a mild and climate-altered winter but it is still winter. And it is unlikely now that we will get out before dark. Right after he says this the wind fires a warning shot. It blasts over the mountains as the sun dips behind some clouds that look swollen and bruised, pink and purple, as they move rapidly from west to east. It is time to go but I am not ready to go. Not yet. I walk downstream to see more of the river. When I crunch across the snow, shadows paint my footprints blue.

It all seems so improbable. That this ice corridor running through this meadow is what provides much of the American West with its drinking water. Thirty-six million people or so in seven states: Arizona, California, Colorado, Nevada, New Mexico, Utah, and Wyoming. Hard to believe it all starts here. Right below this silvery layer of ice flows the blood of the West.

Wallace Stegner asked the essential question: how to make an arid region that is in large part desert or semi-desert into a livable space? What I am looking at right now has long been the answer to that question. And it has been a reliable source. Until now.

Stegner also said that the West is a stage for the heroic. Or something to that effect. I'm not claiming that our little hike should earn us any medals for valor. Maybe if we'd done it in a blizzard. But while we, the actors, might not be heroes, the stage itself is, to use the old name for this river, grand. The white peaks above us, the rushing water, the crystalline air. How perfect that our overly virtual civilization is built on something so starkly real. So vibrantly unvirtual. So purely primal.

Yes, I know that what this overengineered river goes through over the next 1,700 miles or so is decidedly unwild. Yes, I know that the rights to it are so doled out, argued over, and divvied up that states have gone to war over them and that a whole branch of law has sprung up around the question of who has rights to what. Yes, I know all this, and understand that my subjective take, colored perhaps by nibbling those mushroom stems back at the truck, means little compared to all those other uses and needs. But right now, staring at the river that is working its way over and under the ice, I can't help but feel a kind of joy. One simple fact about what I am witnessing here astounds: *Nature is the source.*

I know of nothing more inspiring than this. Nature is the source. We, in our twisted arrogance, are blinded to that simple truth.

I don't believe that nature provides the answer to climate change. But I believe it is part of the answer.

How wonderful, and almost absurd, that so much depends on snowmelt. Snowmelt! Of course, that also means that so much depends on snow, and snow piling up as snowpack. A snowpack that, in its time-release fashion, keeps feeding the river in spring and summer. Snow, which has gone missing this year.

This is the great bounty we have been given. This is the great bounty we have squandered.

But I'm not going to let myself go there right now. No jeremiads or lectures. Instead, I am going to revel in this primal place. Instead, I am going to enjoy being so close to the source.

—

Three months after my western trip, on my birthday, March 15, a year after my travels started, an ice shelf in Antarctica nearly the size of Los Angeles will disintegrate. It will do so after a period of extraordinary warmth, more than forty degrees Celsius higher than normal, on the continent. The Conger ice shelf will be approximately 460 square miles.

The glaciers continue to melt. How we feel about them seems to have little effect on this fact. We ignore them; they ignore us.

As COVID and climate have become more intertwined, so have the personal and the political. I don't think I am alone in this. Deaths of those close to us. Death of the planet. Out of necessity, many of us have become masters of repression.

We repress to live. The end of the world can barely hold our attention. The world is crying out to us but we are not listening. The world is saying, "I have changed. I am different. I am hurting. Don't you see? Can't you notice?" I'm busy, we reply.

We learn to "be adults." To callus over. We learn not to panic. We learn to stay calm despite the circumstances.

But every now and then something breaks through.

—

For me it happens early on the same morning that Mark Karger and I drove up to the source. I skipped that early morning in these notes but let me back up.

The night before our trip to the source, I find myself sleeping up the canyon in Boulder's Adventure Inn, a place where I have never stayed before. I wake to a spectacular sunrise and take my coffee out to the picnic table in front of my room and write for a while. Blazing orange patches of clouds light up the morning sky. I wonder, not for the first time, "Why don't I live here?"

Half as a joke, I use my phone to play "Rocky Mountain High" in the car as I drive down into Boulder. I tear up a little, the silly, sentimental kind of tears. I have no idea real tears are on their way.

My destination is Karger's house outside of Denver, and I listen to "Rocky Mountain High" twice before letting the playlist move on to other songs. I think nothing of it. "Thank God I'm a Country Boy" follows "Take Me Home, Country Roads." And then it happens.

"Annie's Song" comes on. "Annie's Song," which just happens to be, for no particular reason, the favorite song of Barbara Gessner. The same Barbara Gessner who gave birth to me sixty years ago and who is now sitting in a nursing home in North Carolina scribbling down what look like ancient runes, not even decipherable to her, in the same sort of wire-ring notebook she has made things-to-do lists in for half a century or so. The same Barbara Gessner who doesn't really believe that she is in assisted living but thinks she is either in a prison or some scary prep school where people plot against her and where each night "they" take her to a place called "the windowless house," and the same woman who lives in a state of constant agitation, as if there is something to do that, if she could only remember what it was and do it, would solve everything that is wrong with her but that she can't do because she can't quite remember what that thing is, like an itch that can't be scratched, and whose mind and body bear little resemblance to the woman who danced in high heels on top of pianos and who was always the youngest, coolest, prettiest mom, and who loved her firstborn son with all her heart and signed her letters to him "Your Ever-Lovin' Mom." Yes, that Barbara Gessner.

That son of hers has built up some pretty thick calluses over his sixty years, what with lots of other deaths and trials, and though he knows that what he is witnessing is a tragedy, an everyday tragedy but a tragedy still, he never really gets too emotional about it. He deals with her, the problem of her, as if she were any other problem on his very own things-to-do list. He can quite coldly say to his sister, "It would be better if she died," and mean it. He can listen to his wife talk about how awful and sad what has happened is, and nod numbly and agree and say nothing.

Until those first notes of "Annie's Song." At that moment the great stoic disappears. Suddenly he—let's cut the crap, not he but ME—is bawling like a toddler. Suddenly my chest is throbbing and tears are flowing and it won't stop, carrying on right through "Sunshine on My Shoulders" and "Leaving on a Jet Plane." And still I keep crying, a few miles short of Karger's house where I pull over at a neighborhood park. "My poor mom," I wail into the rental car. "My beautiful mom." "*My* mom." "How can this be?"

It is an old-school catharsis and it all comes pouring out.

It takes about half an hour to pull myself together. To stuff the tragedy of my mother back down where it belongs. To carry on with my day. To get "back to normal."

A few hours later Mark Karger and I will be walking alongside the Colorado as we hike toward the source. But there are long stretches where we won't be able to see the river at all.

It flows beneath the ice.

THE GREEN BELOW

This morning I made the long drive down from Denver to the Durango area. I came the boring way, straight south on 25 before cutting west, because snow was threatening, and even then I had to deal with a half-snow, half-rain blizzard coming over Wolf Creek Pass on Route 160. It's now December 7, two days after our hike into the source of the Colorado, and this was my first snow sighting, at least of the coming-down-out-of-the-sky variety, on the whole trip. Not long after the pass I drove through Pagosa Springs and arrived at my destination. Last summer I tried to visit Chimney Rock National Monument but the gate was locked. So today, six hours ago, back in Boulder, I checked the website to make sure it was open. The website claimed it was but the gate says otherwise. But this time I will not be denied. I park outside the gate and hike in.

There is a lot of worry about the overcrowding of nearby Bears Ears National Monument, due to all the media attention, and no doubt that worry is justified. My experience in Bears Ears, however, and in most of the national monuments I visit, is the opposite. I usually see no one or next to no one. This likely has to do with when I visit. For some reason I have always headed to the desert in summer, when other, wiser people visit cooler climes. I also don't mind cold and have recently taken to exploring the Southwest in the winter, when it seems most people would rather be warm or at least skiing. This wrong-footing of the season has allowed me to visit many supposedly crowded places all by myself.

Today I do run into one other pair of humans, a couple who, like me, had to park outside the gate and are wandering aimlessly near the locked visitor's center, wondering how to get up to Chimney Rock itself. They wander off on a trail that I already know goes nowhere. When they circle back I tell them what I think: that the road behind the center leads up to a trailhead that leads to the ancient structures that we have come to see. The three of us set out together up the dirt road. He is from England, she from India, and we chat a bit. They are nice enough, but I want to be alone so say a friendly goodbye and start walking faster. We are at about seven thousand feet, but eight days in the West have left me in pretty good shape. Also, I've nibbled on some of the mushroom stems that Mark Karger left me after our trip to the source. Not too much; just enough to give the day some color, heft, and, possibly, mythos. In fact, come to think of it, I nibbled on a few stems yesterday when I hiked up above Eldorado Springs. *Is this micro-dosing?* I wonder. *Hey, maybe I am micro-dosing.*

Three miles or so later I am up top. And I do mean up top. As in top of the world. There stand the two spires of Chimney Rock jutting into the sky, with a half dozen ravens slowly circling, and there is the panorama of mountains in every direction for hundreds of miles. I reach the trailhead where I would have parked if the gate had been open, which makes me happy it was closed. Earning the elevation is better. The last bit of the hike is short but spectacular, the kind of land bridge that sometimes makes me nervous. But today I walk across relatively unafraid. I'm not sure if the medical community would agree, but I wonder if it is possible that the stems I consumed have helped ease my fear of heights. What I find at the other end of the short trail is astounding, and I will describe it shortly. But while I am exploring, the couple catches up to me and begin *oohh*-ing and *ahh*-ing over the same sight. I know I am being picky here, and maybe a little snobbish, but I really liked having the place to myself. So I decide to wait them out, which doesn't take long. The sun is getting low in the western sky, just above the mountains, or one set of mountains, and they wisely start heading back. I, less wisely, stay.

So here I am. And where is here? I'm not sure I can describe it well enough to make you understand, but I'll try. First, the facts. The Chimney Rock site is a series of relatively well-preserved Ancestral Puebloan structures that ride the back of a ridge that juts nakedly into the sky with views of mountains in all directions and a direct view of the massive upshot of the rock that gives the site its name. The site is a spectacular series of kivas and great houses built upon a narrow spine of rock. It is one of the outlying structures from the great days of Chacoan glory, about 130 miles northeast from Chaco itself, and, at eight thousand feet, is the highest of those structures. It is also, for my money, one of the most beautiful, in no small part due to the view.

I'm glad I didn't see any pictures of this place before I came. You hike up to this high ridge and then all of a sudden you see a wall of chinked stone. Just a remnant wall, you think at first, but then it extends for a hundred feet or so, running straight and true almost a thousand years after it was built. The outer wall is tightly packed yellow, gold, and orange sandstone, spiced with bright orange lichen. The sandstone, obviously, must have been brought from down below, which merits an epic poem in itself. Inside are a series of small rooms and not-so-small kivas, including a great kiva. I peer down into the kivas, perfectly round, where sage and rabbitbrush grow amid patches of snow.

I asked my friend, the writer Craig Childs, to join me on my trip here. He couldn't come so instead I brought his book, which, no offense to him, suffices. I'm not saying it is an adequate replacement, but it's good company. According to *House of Rain*, these rooms I'm looking at housed about 360 residents, or about eighty-one households, in the eleventh century. Which is insane, by the way. This was not a place where you let the kids out to play in the yard. Walk too far out the front door and you fall about a thousand feet. Same for the back door.

An obvious question comes to mind. Why the hell build in such an inaccessible, and frankly, scary place?

One answer is that this was not likely a place to hang out or raise the kids or even to send warriors who could defend the far reaches of Chaco (though it might have been that too) but was an observatory, or, as Craig puts it, "a giant moon clock." Though I have never been here before, I visited Chimney Rock just last night in Craig's pages. In *House of Rain*, he writes: "For various convincing reasons, this site is thought to have been a lunar observatory for the Anasazi, a place that may have been used to confirm and commemorate one aspect of the lunar cycle. When the moon reaches its northernmost point, an event that begins every 18.6 years in the lunar standstill cycle, it rises directly through a pair of massive, natural towers off the end of this ridge."

A thin fingernail clipping of a moon is rising. Why not spend the night? Solstice is close after all, and hey, for all I know, so is the 18.6-year cycle. The wise, middle-aged part of my brain is suggesting it might be smart to leave this knife blade of rock before darkness comes. Both Craig and the remnant mushrooms argue otherwise.

I have a not particularly profound thought. *People lived here. Now they don't.* They built well. They worked hard. They studied the moon. And they left something beautiful behind.

Of course I know nothing gold can stay. Ozymandias and all that. It's all fleeting in a way that none of us, even the poets, can ever quite bring ourselves to believe. Still, why not try to leave a little something, especially if that something blends art and science? This place feels like a message sent through time. I'm not sure I have ever seen the past so surely embodied. A thousand years, slowly passing, on display.

We don't really believe that our own civilization can collapse. *I* don't really believe it. It is a great failure of imagination. The case for our demise is fairly strong: the entire history of the world argues for it. The rise and fall of every other civilization that has ever existed. But still. It can't really happen to us. Can it?

I try to project ahead. I ask scientists to do the same. But maybe we, caught in the amber of time, are not up for the task. Maybe we are tiny flecks of foam in a great river. To think that a fleck can control or even predict the course of a river is a little silly.

In *The Future We Choose*, Christiana Figueres and Tom Rivett-Carnac, the architects of the 2015 Paris Agreement, argue for a "stubborn optimism" when looking to the future, and they believe there is a path forward. This is heartening, but the part of their book I find most compelling is their vision of the future in 2050 if we *don't* take adequate action and the world warms by three degrees Fahrenheit. That date is right around the corner and Hadley will only be in her late forties. The West will be on fire of course, it already is, but since this trip has focused on that other element, let me quote their take on where we will find ourselves with regard to water if we don't make dramatic changes:

> The millions who depended on the Himalayan, Alpine, and Andean glaciers to regulate water availability throughout the year are in a constant state of emergency: there is little snow turning to ice atop the mountains in the winter, so there is not more gradual melting for the spring and summer...
>
> Even in some parts of the United States, there are fiery conflicts over water, battles between the rich who will pay for as much water as they want and everyone else demanding access to the life-enabling resource. The taps in nearly all public facilities are locked, and those in restrooms are coin-operated. At a federal level, Con-

gress is in an uproar over water distribution: states with less water demand what they see as their fair share from states that have more. Government leaders have been stymied on the issue for years, and with every passing month the Colorado River and the Rio Grande shrink further. Looming on the horizon are conflicts with Mexico, no longer able to guarantee deliveries of water from the depleted Rio Conchos and Rio Grande. Similar disputes have arisen in Peru, China, Russia, and many other countries.

Can this really happen? Will this really happen? *We'll see* is the only adult answer. But standing here amid a civilization that was and now isn't, I can picture if not quite feel it.

Chimney Rock is a hard place to leave. And I sense Craig shaming me when I head down before the sun does. Still, it is hard to feel too bad. The dying sun lights up not just the two rocks but the junipers along the road. They become golden torches and their shadows carve up the land, shadows that extend and elongate. It's a magic show that happens every day: the quotidian magnificence, the usual greatness, of a mountain sunset reflecting up red and rippling into the clouds. My legs feel strong, too, and I wish I could stay in the West all winter. A flicker flies across my path. I enjoy the first pulsing of cold.

Saturn peeks out of the blue clouds. If I had stayed up above I would have gotten to see not just Saturn and the waxing moon but Venus and Jupiter, all on display tonight. But I don't mind. I feel like I'm leaving on very good terms with Chimney Rock. Sometimes it's better when a place is closed.

—

Last night I got a hotel in Gallup after stopping in the town of Thoreau, where Louis Williams grew up. The town's Navajo name is Dlooyazhi, which means "Place of the Prairie Dogs." I texted Louis a picture of myself in front of an old building with the word *Thoreau* on it. He told me that on Friday nights the building used to be a drive-through liquor spot at Johnnie's Inn, and that after stopping there they would head up to Castle Rock, the red rock that sits north of the town, to bask under the stars and moon, which lit up the Zuni Mountains on the southern horizon.

This morning I head through the mountains back to the river. Or at least something like the river. This won't be a great elemental reunion where I ecstatically quote Thoreau or Whitman. What I see will be the river impounded. The river caged.

I am heading to Phoenix, which currently holds Chaco's old title as the center of the desert Southwest, with roughly 4,845,832 human beings—*4,845,832*, get your head around that—or twice the population of Denver and thrice that of Salt Lake, and which reached a nice and toasty 115 degrees this summer. Despite this, and the fact that there are now around 110 days each year when it hits 100 degrees, people are flocking there: the population has grown 14 percent over the last decade. I'm sure you can guess my overall take on this place, but before I rail on the city, let me praise it. Coming down the Beeline Highway out of the mountains, I see my first saguaro and am struck by the sheer beauty of the landscape, and for a moment, on this perfect and perfectly warm winter day, it really does earn its name of Valley of the Sun. I even, descending through the Salt River Pima–Maricopa Indian Community, catch a glimpse of light glinting up in the mountains, and see what looks like a line of reflective glass but is in fact exactly what I have come here to see. A canal. Funneling the water of the Salt River down to the thirsty city.

Soon I am driving along the same water, the Arizona Canal, straight and true, and entering the suburb of Scottsdale, which alone has twice as many people as Boulder, Colorado. Again, I need to check my prejudices. Why do I frown on this canal when I marvel at the construction in the nineteenth century of the Grand Ditch, or the irrigation of Phoenix itself, a couple millennia ago, by the Hohokam people? That is one of the many questions I will be asking the archaeologist Ralph Burrillo, who, despite a deep fondness for his natural home in northern Arizona and a lingering suspicion of (if grudging fondness for) Phoenix, currently resides in Scottsdale.

I pick up Ralph at his apartment and he brings me out to two sites that are pressingly relevant to my current speculations. The first is only a mile from his house, a field of cotton, on Salt River Pima–Maricopa Indian land. He explains that while some of the cotton is grown by Indigenous farmers, most of it is leased from the tribes by big agribusiness firms that depend on irrigation and are increasingly subsidized by the government as cotton prices fall. In other words, the government pays for water for an unneeded crop. Unneeded but important, at least in the watery scheme of things, since the purchase of the water by cotton farmers was one of the ways that the staggering cost of the canal that brings the water here was first justified. The canal in question is not the Arizona Canal that I drove along but the Central Arizona Project, the most expensive aqueduct system in the United States, which transports water from the Colorado River, on the state's western border, all the way to this city. The bill for this project was $4 billion, paid for through a government loan, and the loan was to be repaid in part by providing water to the cotton farmers, who at the time were riding high. Until global competition and a pesky megadrought hit.

When we drive away from the cotton fields a couple of reservation dogs come running after our car. Not too long after we see a dead dog on the side of the road.

Ralph tells me that pima cotton, as it is called, was also a prime crop for the Hohokam people for a couple thousand years, and then, as now, it was supported by irrigation. But, Ralph explains, if the drought continues—and who in their right mind doesn't think it will continue?—and prices keep falling, the leaseholders could pull out and the Native owners could be ruined.

Did I say Chaco was the center of life in the Southwest? Well, it was *one* center. This was another. The Hohokam people irrigated the city that stood where Phoenix now rises with water from the Salt River, the same water I saw glinting up in the mountains as I drove into town, and the Salt still provides almost 60 percent of the water that courses through the city in a series of canals. The Hohokam were geniuses of irrigation, and not far from the cotton fields Ralph shows me the ghostly remnants of one of their canals. A dusty path that once, hundreds of years ago, sated the city's thirst. The Hohokam's engineering anticipated the irrigation of modern Phoenix, though they lacked the muscle of electricity.

The canals that weave through this city, Ralph explains, travel atop the dry ghosts of Hohokam canals. In fact, Ralph has moved here specifically to help the city contend with its ghosts. The more the new city builds, the more it uncovers the old. In his role as a cultural resource manager, Ralph helps recognize, and if necessary relocate, the ancient burial sites and graves that are everywhere below the city.

The old city existed for 1,500 years, reaching its peak at roughly the same time as Chaco and falling apart a thousand years ago. Why? Well, at this point, we can list the usual Chacoan suspects. Climatic change, when the reliable Salt River became less so. Too little water and then, in the form of floods, too much. The final blow being the Great Drought that hit in about 1200. Too many people, too few resources. Societal breakdown. As I said, the usual.

It doesn't take a genius to make the jump to modern Phoenix. In fact it is a really hard jump not to make.

But Ralph cautions me about making that jump too easily.

"For one thing, the civilization didn't 'disappear,'" he says.

Later today he will send me a chapter from his own book *The Backwoods of Everywhere*, which elaborates on this point:

Other authors have noted and made much about how most of the canal systems of modern Phoenix are built atop those of the "failed" Hohokam civilization, pointing out how their ambitious irrigation and subsequently explosive birth rate exceeded the carrying capacity of the land until they had to abandon it. Ideas like this are often trotted out alongside pithy observations about how "we" need to learn from "their" past.

The fact is, most ancestral Southwest peoples just weren't as sedentary as most people in the world are today, partly because they didn't have anything even close to our private-property obsessions. Americans will stay in a house that's burning, flooding, and falling apart all at once because they own it—where else are they supposed to go? People like those of ancient Sonora, by contrast, wouldn't think twice about leaving a place when the weather turned lousy and then returning a few years, decades, or generations later when conditions were good again. Why not? It's not like they'd be spending the intervening period squatting on someone else's property, because no property was anyone's property.

Heeding his advice, I will try not to overplay the end of civilization card. (Though, I hear you saying, it's too late for that.) But after a summer where the temperatures recorded were even hotter than during that modern fable known as the Dustbowl, and at a time when I just saw with my own eyes the low and withering Lake Powell, I think I can be forgiven for using the Hohokam not just as a metaphor or parable but as a kind of template for what might happen to this desert city.

From the Salt River Pima–Maricopa Indian land, we drive to one of the canals that run through the modern city, laid like a grid over the ancient canals.

We park in a shopping plaza, ignoring the sign that says *German Sausage Parking Only.*

"Well, we'll get a German sausage later," says Ralph.

We walk along the canal. By this point in the trip I am so tired that I retain little of what Ralph says. I do remember that he tells me that theories in archaeology have a shelf life of about ten years. That is helpful. The theory du jour may not be the theory of tomorrow and definitely was not the theory of yesterday. Time moves. Theories change. The past is not changed, just ideas of it.

Later, when I get back to North Carolina, I will read Ralph's book, in which he writes:

"I love Phoenix for those canals. It's like a mad, dystopian Venice."

His fine chapter on the city ends: "Above all, though, I love Phoenix for the fact that it's doomed. Completely and utterly doomed. I find a sort of resigned comfort in that notion. It's the only place I've ever lived where being naïve about humanity's future is a genuine challenge, and the conservationist in me cannot get enough of that."

After I drop Ralph off at his apartment I go in search of the Colorado River, or what is left of it, Phoenix's portion. This year will be the very first when the federal government announces significant cuts to Colorado River water apportionments, citing the fact that we are technically in year twenty-two of a prolonged drought and even daring to use the phrase *climate change* in their report. Arizona will take the biggest hit, with an 18 percent reduction. (By the time I am typing this, the reduction has been upped to 21 percent.)

The river I am looking for is not the same Colorado I saw below the ice back in the Rockies. To find it I turn where my people turn for such knowledge. To my phone. It tells me that twenty minutes away is the Central Arizona Project building, where I manage to talk my way past the guard at the gate. While the waters of the Salt still flow down from the nearby mountains, the Colorado takes a much longer trip, a trip which requires that its waters be lifted and powered, much of the electricity provided by the dams along the river itself.

The receptionist at the CAP center, who just started last week, doesn't know where I can see the canal, but she does hand me a sheet of paper that indicates certain points where it might be possible. Before leaving I sneak out back to see if I can catch a glimpse of the canal, but the vibe is vaguely military, and, feeling paranoid, I retreat.

Maybe I'm just tired. Ten days of mostly following the river have worn me down. I have a strange worry that I have a blood clot in my hand. Can people get blood clots in their hands? Probably I just bruised it. I remember two falls. One in Cathedral Wash. The other on the hike into the source. These seem more likely suspects than a sudden hand clot. I find a promising sign ten minutes from the CAP center. I park at something called Reach 11. I am fried and walk for about a mile in the dust. Thunder rumbles. I quit and turn around. No river. A failed quest.

I am ready to give up and leave Phoenix behind, but about a mile from Reach 11 I drive over a bridge and see water flowing below. A sharp, squealing right turn brings me into a vacant lot next to a chain-link fence. Concertina wire, like that I saw in DC on the fences surrounding the Capitol building, tops the fence. Almond Joy wrappers and empty Sprite bottles litter the ground, but on the other side of the fence I see it. The once-mighty Colorado, or at least some remnant of it, here tamed and small. It's like seeing a mountain lion in a zoo. This is imprisoned water, forced to do our bidding, in a canal between banks of dirt by the highway, the cars loud. It flows straight with purpose. Coming soon to a faucet near you.

"What to make of a diminished thing?" asked Robert Frost. This is the diminished river and I say goodbye. I will not be pulling on my loincloth, climbing the fence, and immersing myself. There will be no elemental ceremony today.

Instead I drive thirty minutes to Connolly's Sports Grill off of the highway. I have come to see someone I have never met before. The brother of Mark Honerkamp, my usual traveling companion. For an hour or so we do what all men do when they don't know each other but have an absent friend in common. We tell stories about (and mock) the one who is not with us. We bond happily over a beer or two.

After a hearty dinner of shepherd's pie, I head north. The final leg. Uphill, as it were, to Flagstaff. The sky darkens as I drive and then I begin to notice something unusual falling out of the sky. Something I have just seen once before during my ten-day trip out west in December.

Snow. I pull over in a rest area in the mountains and let it land on my face. If not a blessing, it is close enough in this dry land. I am not a religious man and as a rule I don't pray. But I pause and say a silent prayer for this world.

—

I am scheduled to fly back from Flagstaff the next day, but my flight will be canceled due to snow, which I will take as a hopeful sign. But when I post a social media complaint about my delay, the first reply will be, "At least it wasn't a tornado." That will be the first I hear about the deadly December tornadoes in the Midwest. Back home in North Carolina the seasons will continue to be out of whack. For ten days after I get back it will be in the seventies, sometimes eighty. Bathing-suit weather. People everywhere joking about the end-times.

No wonder conspiracy theories and lies are so popular these days. No one knows what is going on. We say these spouters of lies are fools and they probably are. But they are also desperate people acting desperately.

We are hitting the hard wall of the future.

Since I started working on this book, barely a week has passed that hasn't offered up some harbinger of a changed world. It wasn't just drought that did in Chaco. Societal disintegration and disease helped too. We have entered an erratic time when only global warming itself, unlike so much else in today's world, is proving reliable.

The hits will keep on coming. A new variant of the virus has bloomed. It is sixty-seven degrees in Alaska in late December, and that isn't even the scariest environmental news. There has been more movement below a glacier in Antarctica. Winter fires in the West have been followed by a rash of flash floods. Meanwhile the world is at war, which means our environmental future, always a low priority, takes a back seat. In the midst of this, and skyrocketing prices at the pump, oil companies report record quarterly profits. Billions and billions of dollars.

With every passing day Orrin Pilkey starts to seem like nothing more than a realist. Understated, even.

In early May I visit Orrin at his retirement home. We talk about heading back to Topsail, but our adventures have grown smaller: we settle on eating lunch near his apartment. Orrin is eighty-seven and tired. Saving the world is hard work.

At the next king tide I drive up to Topsail in Orrin's honor and watch the water lap against the sandbags and the base of apartments and houses. "I don't believe your beach has a chance of lasting," Orrin said to the homeowners more than a decade ago, but the beach, thanks to the dumping of sand from elsewhere, is still here. Not that Orrin has been entirely wrong. A whole row of houses has gone in the drink. But Topsail has just rebuilt and dumped sand. That that sand will go away soon enough doesn't deter the citizens, and if we are honest it likely wouldn't deter you or me if our houses depended on it.

That is the realist in me. The person who doesn't really believe that I, or my words, can change the course of where we are heading. And yet there is another me who wants to yell in warning and shake my fist and let loose with a jeremiad.

That part of me wants to ask: don't the executives of these fossil fuel companies understand, or care, that future generations will revile them? Apparently not. Apparently they lack the empathy and imagination to see that they will be clumped with other historic villains: perpetuators of atrocities and holocausts, slave traders and owners. While I imagine these people must have substantial egos, they seem to be unconcerned with the opinions of others, especially the imagined others of a vague future. But what they have done will be remembered. Their grandchildren will look back with shame. Their names should live in infamy. Their names *will* live in infamy.

And so we find ourselves, a quarter of the way through the twenty-first century, on a hurtling plane flown by arrogant rich men beholden to the whims of a dying industry, who are trying to squeeze out the last pennies while destroying the planet in the process. And where are they flying this plane? Into a volcano. Imagine the anger we would feel if we really thought about this. If we let ourselves imagine and feel.

—

Eight months after my December trip I fly back to Flagstaff. We get within twenty miles of the airport when thunderstorms drive us back to Phoenix. It is 116 in Phoenix, but it has also been a rainy summer, a strong monsoon season. So rainy that when I reach out to Ralph Burrillo and ask him my question about forty-two years, his answer is a contrarian one.

"No two climate experts, whether real or merely 'experts' in their own minds, can agree where southern Arizona is headed," he writes.

He admits that the water is dwindling in lakes, rivers, reservoirs, and aquifers, which "suggests this place will be an arid wasteland by forty-two years from now." But then he throws in his wrench, the monsoon cycle.

He points out that it has rained either in Phoenix or nearby every single day for the last month. "In brief, the monsoon cycle is driven by heat—hot air rises as the ground surface is heated, creating a vacuum effect that 'sucks' in moist air from the Pacific and the Gulf, which take a lot longer to heat up. And the more energy there is in that system, the more intensely it operates. So, somewhat ironically, the more solar heat beats on southern Arizona the more it exacerbates the monsoon cycle, at least in theory." He says that the faucets in Phoenix will be running dry, "but at the same time, that intensification of temperature could—and, at least this year and last, very much *has*—resulted in intensification of the monsoon cycle to such an extent that this place becomes a lightning-assaulted jungle for at least part of the year."

He continues, "Bottom line, as much as it pains me to say this, southern Arizona might become a haven in the coming years, at least if the lightning doesn't get ya. The rivers will dry but the rains will fall. And I guess that also makes sound ecological sense. Climate change, like all ecological phenomena, can't be called a Good or Bad thing altogether because what's awful for one niche can be equally beneficial for others. Like, I personally think killing off America's wolves was an egregious bunch of amoral horseshit, but you won't hear any complaints from the coyotes that rushed to fill the void. They're fine with it."

Picturing Phoenix as jungle certainly scuttles the usual apocalyptic visions of the place. I am reminded of something Orrin Pilkey said: "The point I'm trying to make is not that *I* can predict what is going to happen. It's that no one can."

—

After finally making it to Flagstaff, I am sitting in the hotel bar when my phone alarm goes off. Its song is echoed by the song of a dozen other phones in the bar, like the courting of spring peepers. What our phones are warning us about is a rash of local flashfloods, floods that have been ripping down the burn scars outside of town. I think again that if your subject is disaster you don't have to work very hard to find it these days.

I have returned west to attend the annual celebration thrown by the Navajo organization Utah Diné Bikéyah in the meadow below the Bears Ears in southern Utah. But I am also here to visit Lake Powell in Page, Utah. "Founded 1957," the sign announces when I enter Page, which leads me to wonder how much longer it will last. That is because the thing the city was built on, the whole reason it is here, is going away. Thanks to some recent snowmelt, the lake is actually seven feet higher than it was when I flew over it with Chris in December, but it is still 161 feet down from capacity. The water in the country's second-largest reservoir, the city's lifeblood, is withdrawing, withering, the green-blue puddle, always incongruous in the middle of the desert, falling. And if the water goes, if it really falls far enough, the thing that blocked the water's flow, that created the lake, will be rendered useless. Glen Canyon Dam will stand like a 583-foot-tall ruin, a symbol of the apogee of the optimism and arrogance of its time. Back when we thought we could build ourselves out of anything.

For the fifty-six years of its existence, the Glen Canyon Dam has been a magnet of controversy, in no small part because building it drowned the beautiful ecosystems of Glen Canyon. In *The Monkey Wrench Gang*, the writer Ed Abbey famously suggested a rather direct way to restore those beautiful canyons: by blowing up the dam. But now it isn't just monkey wrenchers who are seriously talking about closing down the dam, and the drought appears to be doing what explosives did not. With Lake Mead operating well below capacity and Lake Powell operating at less than 25 percent, there is serious talk of shutting it down and letting the river run unimpeded from here to the Grand Canyon. And if that happens, who would bet against Page becoming a ruin or at least a shell of its present self?

I decide to part with some cash and pay for a spot on a tour boat to see the newly reduced lake for myself. There are four of us in the small boat and the tour guide, with a kind of Jimmy Buffett affect, wants nothing of my environmental blather. As for the annoying couple from Michigan, they make their politics clear enough when the man, staring up at the canyon walls, says: "If there really was global warming this lake would be full. Everything would be melting." I don't take the bait and point out that there are no nearby ice caps.

Otherwise it is an enjoyable trip. We putter out from the launch past rows of docked houseboats. Our captain points out all the new landforms that have emerged over the last couple of years and how navigating the lake is an ever-changing adventure. Soon we are staring up at sandstone walls that rise 220 feet above the water. The lake, like the rest of the West, is at low tide.

We pull up to the dam itself, which has an industrial look reflective of the time it was built. Both utilitarian and military, a practical fortress.

Bashing the dam is a popular sport for environmentalists and I am far from immune. But I am also aware of something going on that, like Ralph Burrillo's take on Phoenix, complicates the picture of the falling water. My plane trip with Chris in December allowed me to witness the landscape of sludge and mud that the withering reservoir had created below Cataract Canyon. This is to the north of the reservoir where the river flowed in, near the delta. As the water fell it revealed all the muck that had built up during the years the dam had slowed its flow. But at the time I was not aware of something else the dwindling lake is creating. And this something is, however much I may not like the pesky word, hopeful.

"My experience with Lake Powell prior to this last year was as a river runner coming down Cataract Canyon," Zak Podmore, the writer and environmentalist, told me when I visited him in Bluff, Utah. "That was a complete disaster zone. Where all the mud from the Colorado River has accumulated. Two-hundred-foot banks of mud under the water and above it. It smells terrible and you can't get out of your boat anywhere. And that is what I expected to find below the delta zone, too. Down in Glen Canyon."

But that is not what Zak has found over the last six months as he has explored, working on his book about Glen Canyon's comeback.

"I found a completely different world of ecological restoration in the hundreds of side canyons of Glen Canyon. If you get ten miles below where the river once flowed, below the delta zone of sludge, there is very little sediment that has accumulated so that there are nice beaches, instead of sticky mud, and pretty much every side canyon has a flowing stream in it. There are willow groves that have grown back that over the last few years have grown eight feet tall."

What Zak has found is a remarkable diversity of native plants, wildflowers, grasses. Tumbleweeds blow along the beaches, but not far inland native plants grow and, within a fifteen-minute walk, you find idyllic groves of cottonwood that are sprouting, willow, beavers damming the creeks in a lot of the canyons, native bees buzzing around. Zak mentions that he had heard of an ecologist who was walking in the canyon for the first time who said it was the most intact native ecosystem he had ever seen despite the fact that it had been underwater just two or three years ago.

"It's not really clear to me why the restoration is happening like it is but it indicates that if you give nature a chance it will restore itself. There hasn't been any active management on the part of scientists or the Park Service, the plants and animals are just returning by themselves."

The monsoons, the rains that come hard this time of year in the Southwest, have helped scour the sludge out of the canyons. When I think of this comeback of the natural world I can't help but think of Ryan Lambert and what he is trying to do down in the Gulf. The idea of nature as restorative. But as in the Gulf the actions of humans can either nudge the natural world in a restorative direction or, as we so often do, get in the way.

"There are going to be impacts to Glen Canyon for a thousand years if the dam comes down. The big question is going to be what's going to happen in the next couple years. It seems very likely that Lake Powell can't be sustained in the long term. All the climate models show that it is not very likely that Lake Powell and Lake Mead could ever fill again with the current water demands in the Southwest. So Lake Powell is more of a liability than an asset right now. It's not providing any storage because there is all this unused storage capacity at Lake Mead and they are scrambling to keep hydropower going at Glen Canyon Dam. So it seems they have to think about modifying the dam to let the water drain, but a lot hinges on how quickly they do that. If they did it intentionally all at once, all the mud at the top of the river would be flushed out. But if they keep putting off the decisions and let the river go up and down really slowly, then the glaciers of mud are going to keep working their way to the dam over the next decades.

"There is a huge chance for undertaking one of the biggest ecological restoration efforts that has ever been done intentionally in this country."

The canyons that Zak is describing are significantly upstream from where I am floating today. Here, closer to the dam, the water is still 360 feet deep.

We turn away from the dam and head toward Antelope Canyon. We stop and stare up at the walls that rise hundreds of feet above us. Up above the Navajo Sandstone shines orange, occasionally stained black by the desert varnish, but not far down the rock looks bleached. This is the famous bathtub rim, which indicates the old high point of the lake. From that line down, calcium carbonate and other minerals were left behind by the retreating water, leaving the walls an almost-white color that from below at least looks quite beautiful. It is a clear marker of where the water was.

We wend our way into Antelope Canyon, the water still deep at first but getting shallower as the winding walls rise. I take a swim in the clear green water, surface-diving but unable to get close to the bottom. If this water were drained all at once in a great flush, we might begin to see the miracle of recovery that Zak has witnessed further upstream where formerly drowned landmarks are reemerging.

I think of Zak's description of one of the canyons that has come back to life:

> There's a thirty-foot slot canyon in Fifty-Mile Canyon in the Escalante arm that was packed to over the slot with mud and sand until the monsoons came through last summer. Now, you can walk on bedrock through this canyon that is six feet wide and then pinches down to a one-and-a-half-foot winding sandstone slot above you. So you are in this amazing little flowing stream with the light filtering down and no indication that it was once 120 feet underwater.

When we spoke he told me about visiting one of the most famous of those landmarks, Cathedral in the Desert, which until just recently could only be explored by boat.

"The bottom of the chamber of Cathedral of the Desert was out of the water last year for the first time. But it had thirty feet of sand at the bottom. Last summer and fall two monsoons cleared it out so now you can stand on bedrock."

To have died and come back. Nothing stimulates the human brain more than a comeback. A reemergence. Whatever one's thoughts about hope, and the dark future, it is hard not to be taken with this idea. The idea that, near death, near disaster, we might return to life. That what we were sure was lost has been found.

—

I have said before that I am wary of hope. But I am what I mock. Looking back on the dozen books I have written, I see a pattern of hopefulness. The story that was central to my early books was that of the ospreys, birds that were decimated by DDT and then came back after the government outlawed the chemical. I came into the present book determined not to be hopeful. Hope, I understand, can be delusional. I wanted to be a realist. To see this world with cold eyes.

I am not going to turn around now and follow the clichéd path of my soft genre. I'm not going to do it.

But still. While I will not allow myself to end on an uptick, I see strands. Strands that if not hopeful, given the fate my daughter will likely face in forty-two years, contain possibilities. I also see some reassurance in uncertainty. That is one thing that wildness is, isn't it? An unclear path, a path that doesn't follow the models. I will resist braiding these strands into a rope. I will not end inspirationally. But it would be dishonest not to include these threads in my thinking.

There is also a simpler, more practical purpose of hope. "Without hope there is no endeavor," said Samuel Johnson, and I know hope can be fuel for doing.

Maybe it is soft-minded of me to return to hope after all I've seen. Maybe being hopeful is naïve in the face of a heating world. I don't want to make that essayist's turn here at the end. But if I were looking for hope I know where I would find it. I would find it in those slot canyons that Zak described, coming back all on their own. I would find it in nature. The living natural world is where I feel, if not hopeful, then whole. It is the origin, it is the source, not just of the Colorado River, but of so much. So much creativity, so much energy, so much true diversity, so much possibility. And maybe that is the word I want to use here. *Possibility*. Because, strangely, one of the most reassuring things is the fact that we really don't know where we are heading.

A FIELD GUIDE TO EVERYTHING

The drumming has been going on all night. Rain patters on the roof of our tent and thunder occasionally announces the arrival of the lightning that soon illuminates the tent's innards, but the drumming doesn't stop. It is still going when I get up at four to take a leak. The rain has ended but not the beat. There is singing too. It is coming from a tepee about a quarter mile away. I try not to wake my tentmate Mark as I zip myself back into the tent, but I fail. "Maybe they are having a peyote ceremony," he mumbles before falling back into a snoring sleep. They are not, I'm pretty sure, but the drumming and singing continue until six in the morning.

The music ends not long before I climb out of the tent. Though it is July in southern Utah, here at seven thousand feet it is a cool, almost cold, morning. The rain has kept the dust on the road down as I follow it toward the center of camp. Deer wander idly through the campsites, as they have all weekend. Wind blows through the pines.

"Everything is living," one of the Navajo elders, Hank Stevens, who represents the Navajo Nation for the Bears Ears Inter-Tribal Coalition, said to me yesterday. "The ponderosa pines are singing to us. The music of the wind through the pines."

I crave coffee but knowing they likely haven't put it on the stove yet, head out to the meadow. The meadow is a field of sage and ponderosa, and I make for the stump where I like to watch the sunrise. A raven shrieks and chases another raven.

Otherwise it is quiet here, the members of our small settlement of about 150 humans just starting to stir. I take my seat on the stump and stare up.

And there they are. The twin buttes that give this place their name. The morning sun has hit them first and their tops glow orange, the rock made even more radiant by the bulky blue-gray clouds that serve as a backdrop.

You can see the Bears Ears from all over the Four Corners, and I first caught sight of them on this trip before I entered Monument Valley back in Arizona. Now, sitting in the shadow of the two peaks, I think: "Something great has happened here."

Something great is always happening here. For thousands of years it was a meeting place for multiple tribes. In the 1860s it was the place where Chief Manuelito held out against US government forces before the Long Walk. But there is also more recent greatness. Back in December of 2016 something of a miracle occurred in this place. After centuries of tearing land away from Native Americans, the United States took a small redemptive step in the other direction. President Obama, heeding the call of the Bears Ears Inter-Tribal Coalition, a group made up of the Navajo, Hopi, Ute, Zuni, and Ute Mountain Ute Tribes, declared the 1,351,849-acre Bears Ears National Monument. This was land sacred to the tribes, ancestral land whose historical and spiritual significance could not be overstated. And now it would be protected, not taken away, by the United States government.

But not so fast. A year later, that promise was broken (does that ring a historic bell?), the monument eviscerated, reduced by 85 percent, by the new president, Donald John Trump. The tribes, which had been lifted in triumph, were now slammed to the ground. Trump's actions also posed a threat to the very act with which the land was created, the Antiquities Act, which gave presidents the power to declare national monuments.

The declaration of the Bears Ears National Monument by

President Obama in December of 2016 had been a moment of hope. Perhaps unity is impossible in our faction-torn world, and in fact the tension in Utah sometimes rises to the level where you might expect a civil war to break out at any moment. But in at least one way Bears Ears speaks of union, not division. The creation of the monument represents a possible confluence of old-school land preservation ideals with the vision that grew out of the work of the tribes, led by Utah Diné Bikéyah.

During the summer of 2018, after Donald Trump dramatically reduced the national monument that was meant to protect this land, I was lucky enough to join the summer celebration of the tribes in this meadow below the Bears Ears. That summer we felt threatened and, despite ourselves, defeated. Local ranchers moved the signs that directed people to the campsite, tailgated our cars, buzzed us with a plane. This summer has been different. Better.

That is because this past October saw another grand reversal. Poetically, and powerfully, the very first Native American interior secretary, Deb Haaland, recommended restoring the first national monument to fully grow out of the thinking, support, and political power of Native American tribes, and on October 8, 2021, President Joe Biden signed that restoration into law.

I am happy to be back here. Bears Ears is a story of great loss and possible redemption. It is both a thoroughly modern story, one that in its battle of entrenched interests reflects where we are as a country and a culture, and an ancient one where the drama is being played out on an ancestral homeland where people have lived and worshiped for thousands of years.

As a writer, I see something else that Bears Ears can offer. Language. A new story. A story that draws on the present and threads in the old and also the ancient. That takes the best of our conservation legacy and makes it better.

—

Back in December, seven months before the summer celebration, I took a hike with Louis Williams and Greg Lameman, a Navajo river guide whom I met a decade earlier.

The day started with quiet.

Well, actually the day started with a coffee-fueled, anxious, three-and-a-half-hour drive from Page, Arizona, and a harrowing climb up the dirt switchbacks of the Moki Dugway, a road I swore I would never drive again due to my fear of heights but needed to take to cut a half hour off my trip so I could meet Louis at the ranger station in Bears Ears at nine thirty.

Then came quiet.

Real windless silence. The kind you never experience these days. Kind of stunning really. Louis, it turned out, was a little late and so I took a pre-hike down the first section of the Grand Gulch trail. The only sound was my footfall on the orange dirt of the path through the high sage and juniper scrub, and then, when I stopped and stood still and stared up at the twin buttes that gave Bears Ears its name, there was no sound at all.

The silence continued. I noted the lack of the music of birds on this winter day. No animals either. This would continue all day, though I would see many prints and at sunset I would at last catch sight of a skulking coyote and a full-racked buck.

When Louis arrived we hugged hello, virus be damned.

"You look happy, David," Louis said, and I nodded.

We waited for Greg at the pullout to Road Canyon in Louis's sister's truck.

Greg arrived and followed us in on the rutted dirt road, then parked in a place hidden by some junipers and joined us in the truck. I met Greg Lameman over ten years ago, when he was a guide on a rafting trip I took, and met Louis Williams a few years later, when Greg introduced us. But today they were not my guides, or if they were they were unofficially so, and so we celebrated our reunion with a beer. Louis did not join us. I drank an IPA as we jostled further down the trail in the truck, Greg a Pabst Blue Ribbon.

"Hydration," he said.

We parked and began our hike down the drainage. The water from the wash, if there were water, would eventually flow into the Paria and then into the Colorado. What water there was that day was mostly of the solid type, ice covering the puddles. The land, too, was hard from the night's cold, a red winding road that went deeper as the walls rose around us. A sculpted dead juniper marked a turn in the canyon, as if pointing the way. A tiny flower grew out of the stone below the tree and Greg broke a bit of it off, crushed it under his nose, and declared that it smelled just like Icy Hot.

Louis was right. I was happy.

It was a glorious morning and I was reveling in hiking down the dry wash, scrambling around the cryptobiotic soil and coyote scat, amid the piñon pines and twisting junipers, the sky a striking blue and the sun shining. My companions appeared to be feeling equally good. When I stopped to take a piss at the base of a juniper, Greg yelled to Louis: "Dave's taking a Lizzie."

"A thin Lizzie?" Louis punned.

"At my age all my Lizzies are thin," I yelled back.

All went well until lunch.

We stopped and ate on a high rock ledge. Our destination was called the Citadel, which sat at the end of a jutting rock peninsula high above Road Canyon in Cedar Mesa, but while I was aware that we had been hiking toward it all morning I didn't know exactly where the Citadel was. Now I learned that it was in fact on a little rock island in the sky, a blossoming mushroom of rock really, directly across from where we were sitting, and that the only way to get to it was the narrow land bridge in front of us, a fact that Louis casually informed me of as I ate my roast beef sandwich.

"We'll cross the bridge after lunch, David," Louis said, nodding his head in the direction of the stone skyway that looked, from where I sat, no wider than ten feet.

Did I mention I am afraid of heights? As we hiked across, my confidence wavered. My daughter likes to tease me when we stand atop high buildings, but I would not call my fear extreme. I prefer the word *significant*. And varied. I am not afraid of going up in tiny planes for instance. But this was just the kind of thing that gets to me. The walkway looked too narrow.

Who would build a home in a place like this, I wondered, as I began my wobbly walk. More importantly *why*?

As it turns out, the answer to this question is an important one, archaeologically speaking. I was not going to back out of course. Not in front of Greg and Louis. On shaky legs that were now acting their age I made my way slowly across the land bridge. *Don't look down* had been my motto this morning as I drove up the Moki Dugway and it also came in handy here.

I felt a little sick by the time I made it across, then instantly relieved, forgetting I had to cross again on the way back. The payoff at the other end was well worth it: A tightly packed red wall of stone with six windows tucked defensively into a sheer sandstone wall. Ancient wood, likely juniper, framed out the top of the windows. The walls were irregularly patterned with tightly chinked stone.

Whoever lived in this place we now call the Citadel did not share my fear of heights. And whoever lived here wanted to get far, far away from other people.

Scientists believe that the area originally—and now again—saved as Bears Ears could have as many as 250,000 ancient sites.

From our high perch Louis pointed at the walls of Cedar Mesa Sandstone down in the canyon.

"You can see climate change in those walls," he said. "You can see the years of drought and the years of rain in the fluvial layers."

Spending his days walking through these ancient homes gave Louis a different perspective. He understood that civilizations don't last forever.

"They were farmers," Louis said. "Why were they so far away

from a place they could farm? Obviously they were protected here. Very protected. But they were farming elsewhere. There is no water here."

Why consciously build so far from water? Why build in such a precarious spot? What necessitated such defensiveness?

What my archaeologist friends have told me is that this land had been abandoned during "Pax Chaco," when Chaco was at its height and the climate and culture were going strong, but that people came back here when Chaco fell. In this way it makes sense. This was a place people would go when things fell apart. A place people would go when they were afraid.

Is it really so farfetched to think this is the last time this place will be occupied? Not if we expand our perspective and think not in terms of days or decades but centuries and millenia. Then the notion seems quite reasonable. If someone were coming after you, and all hell had broken loose, it would not be a bad place to be. A home with its back to the wall.

—

It is good to be back here, up in the cool above the heat.

We are above the world, at seven thousand feet. Up above it but not beyond it.

This morning, the second of the celebration, I talk to Jessica Wiarda, a young woman only a couple years older than Hadley. Jessica is half white, half Hopi and is the artist-in-residence with the Utah Diné Bikéyah, the Navajo group that was the driving force behind preserving Bears Ears as a national monument.

The Bears Ears themselves loom in the background as she describes growing up in the city of Logan, Utah, where there were no other Native Americans in her class and classmates sometimes called her "Pocahontas." This is her first time in Bears Ears, but she feels connected to the place by stories. She says that her grandmother told her that *her* grandmother told her about her grandpa hunting here. Now that is a generational imagination.

We talk mostly about the place, its beauty, its quiet. But when I ask her about the climate crisis her face changes.

"I guess I try not to think about it. Because I get teary when I do. Especially with Salt Lake drying up. People were mourning about it. So I'm not very hopeful. I'm not."

She pauses, falters, continues.

"I feel like I'm going to be old and dealing with the heat and lack of water. I will have to save more. I'm blessed that I have a job and can save money. But I worry about my friends that are living paycheck to paycheck."

She now cries as she talks. We are both caught by surprise by the sudden emotion.

"It's the first time I've really talked about it. I usually don't want to talk about it. But if you ask me, I am not hopeful. Maybe once things are bad enough, when it's really here, maybe then the generations can work together. But no, I'm not too hopeful."

Later, around lunchtime, I talk to Woody Lee, the executive director of Utah Diné Bikéyah. He sits on my favorite stump in the meadow amid the sage and tells me the Navajo story of the twin warriors who fought the monsters that threatened their people.

"Some of the monsters begged for their lives. They said, 'Put me in a place I cannot be seen again.' And they were told, 'You will be placed in the ground. That's where you will stay. And nobody will let you out.' The monsters agreed but said: 'If someone takes me back to air and light there will be repercussions.'"

He pauses.

"One example of this sort of monster is uranium. That was a monster that was taken out of the ground and look what happened."

Not long ago there were uranium mines, and the attendant cancer deaths, within Bears Ears. A new mine was launched in 2019 when Trump undeclared parts of Bears Ears.

"The way we approach the land is the way we will leave it. If we dig stuff up there will be repercussions. You don't just take things from the ground in any place you want to. When we come to a place we make an offering."

I ask about Trump's reduction.

"When Trump reduced the size of the monument it had an unusual effect. It brought attention to the issue. We gained more support when he reduced the size. We gained even more when he sent Secretary of the Interior Zinke out here. We were gaining momentum, we had more people joining us, more people supporting us. He did not know what he was waking up. He was waking up the bear.

"Bears Ears will continue to be a place we hold up as an inspiration, a place that orients our prayers and a landscape where people from all cultures can come to pray, to heal, and to restore our minds and bodies."

We leave the meadow and Woody heads over to lunch. I walk back to our campsite, by the big tepee where we gathered last night for sensitivity training. I have been through these things before and to be honest was mostly tuned out, but after the basic program three of the Navajo elders started to talk and the experience went from rote to inspired. They spoke about how much Bears Ears meant to them, the devastation of their people by COVID, the importance of being in a place where "it all started."

The line that stuck with me was one by Hank Stevens: "We came into this world and use our words for the world."

I was not sure I had the line exactly right, and when I asked him the next day Hank was not sure either. But it remains an important sentence for me.

Words still matter in this chaotic world. People in Washington D.C. use their words to say this place is saved, then say it is unsaved, and for all we know the old president might soon become the new president and take this away again. But it still feels good to be back here, whatever its designation. It remains a place filled with energy and legend.

Usually those who write the laws for a place are divorced from the place itself. The document declaring Bears Ears a national monument, however, grew out of the original proposal written by the Bears Ears Inter-Tribal Coalition. The final version was written by Interior Secretary Sally Jewell, who had hiked, rafted, and explored the area, and the acting chair of Obama's Council on Environmental Quality, Christy Goldfuss. It began:

> Rising from the center of the southeastern Utah landscape and visible from every direction are twin buttes so distinctive that in each of the native languages of the region their name is the same: Hoon'Naqvut, ShashJáa, Kwiyagatu Nukavachi, Ansh An Lashokdiwe, or "Bears Ears." For hundreds of generations, native peoples lived in the surrounding deep sandstone canyons, desert mesas, and meadow mountaintops, which constitute one of the densest and most significant cultural landscapes in the United States.

But where the proclamation gets really interesting, for me at least, is when it gets particular:

> The diverse vegetation and topography of the Bears Ears area, in turn, support a variety of wildlife species. Mule deer and elk range on the mesas and near canyon heads, which provide crucial habitat for both species. The Cedar Mesa landscape is home to bighorn sheep which were once abundant but still live in Indian Creek, and in the canyons north of the San Juan River. Small mammals such as desert cottontail, black-tailed jackrabbit, prairie dog, Botta's pocket gopher, white-tailed antelope squirrel, Colorado chipmunk, canyon mouse, deer mouse, pinyon mouse, and desert woodrat, as well as Utah's only population of Abert's tassel-eared squir-

rels, find shelter and sustenance in the landscape's canyons and uplands. Rare shrews, including a variant of Merriam's shrew and the dwarf shrew can be found in this area.

Carnivores, including badger, coyote, striped skunk, ringtail, gray fox, bobcat, and the occasional mountain lion, all hunt here, while porcupines use their sharp quills and climbing abilities to escape these predators...

Raptors such as the golden eagle, peregrine falcon, bald eagle, northern harrier, northern goshawk, red-tailed hawk, ferruginous hawk, American kestrel, flammulated owl, and great horned owl hunt their prey on the mesa tops with deadly speed and accuracy. The largest contiguous critical habitat for the threatened Mexican spotted owl is on the Manti-La Sal National Forest. Other bird species found in the area include Merriam's turkey, Williamson's sapsucker, common nighthawk, white-throated swift, ash-throated flycatcher, violet-green swallow, cliff swallow, mourning dove, pinyon jay, sagebrush sparrow, canyon towhee, rock wren, sage thrasher, and the endangered southwestern willow flycatcher.

There is a tendency to over-romanticize what is gone. There is also a tendency to over-romanticize Native people. I want to avoid that but I find in the example of Bears Ears a combination of hardheaded sense—practicality of the sort Ryan Lambert has—and a deep love of nature.

Wildness is not just nature. Wildness is surprise and uncertainty. That is another thing we are destroying. The way we corral language these days. We have become safe, predictable, dull. We hunger for the spontaneous, the surprising, the wild.

—

The theme of the weekend is the living world.

More specifically, the theme is: *Iina bi'gaal*, which translates to "letting life speak."

I am not making any sort of claim that what I find at Bears Ears provides an answer to our crisis-filled world. I make no claim at all. I just know that for me, personally, the theme fits where my thinking is going.

What is to be done? is the question of course, the forever question. Even if I claim to want to sing, not save, the world, I don't want this world to go away. I want Hadley to be able to explore the beaches and canyons and forests just as I have. I want her children and their children to be able to do the same.

Back in January I made some resolutions. I was pretty confident that I'd accomplish a couple of my more practical ones. It is finally time to put up the solar panels and buy an electric car. To at least practice a little of what I preach.

I am aware of how little that will do to forestall this massive thing rushing toward us, the massive thing we are already in. Michael Mann makes this point in his book *The New Climate War*. These personal choices matter, of course they do. But too much focus on them is putting the emphasis on the wrong thing, like telling someone that the way to win World War II is recycling your cans. (Of course recycling *was* part of the war effort.) Worse, this can be used to undermine the one thing that will really change anything: massive governmental action and regulation. That starts with a simple realization: we are in the midst of an existential crisis. Existential as in *exist*. Hello! How can that not be enough to wake us up?

Impotence is more than frustrating. I remember coming out of *An Inconvenient Truth* all those years ago vowing to change the world. But I couldn't change the world. I could barely change myself. About all I can do, really, is write. And I don't have high hopes for that doing much world changing.

Michael Mann also takes exception with the extreme doomsayers, claiming they undermine the efforts to fight. I have always been in that camp, in fact I once wrote a whole book about my suspicions about apocalyptic thinkers, and their kinship with religious end-time thinkers. But I find my resistance to that sort of thinking is softening. When you spend enough time in the apocalypse you start to believe in it.

And yet still. People are all allowed to have their opinions about the world, and we have grown used to hearing them express those opinions with confident adamancy on CNN, but for all my love of Orrin Pilkey I prefer voices that are a little less sure of themselves.

And so my New Year's resolution was not "Change the US government." I decided to leave that to tougher minds than mine. What I came around to was something softer, more personal, something a little like this: I want to *remember this world.* As my mother forgets everything, I want to remember.

In fact, now that my year and a half of travel is ending I am thinking about a new project. In my head I'm calling it *A Field Guide to Everything.* I will learn the birds, the plants, the mushrooms, the trees, the stars. All the animals before they are gone. What better pursuit as the world burns? Fearing that I will follow my mother into forgetfulness and oblivion, I will memorize the world. It will be an effort toward expansion, toward empathy.

Toward humility, I would say, though Nina sees right through this.

"Then you really will be a know-it-all," she says.

I see her point. Historic know-it-alls, like Humboldt and Linnaeus, have tried this sort of thing before. What has it got them?

Naming a semipalmated plover or an oyster mushroom will not stop my mother from forgetting. Will not stop me from forgetting. And will not stop the world from catching fire.

Still, my resolution was strengthened recently when I read a few lines in a book by my teacher from Colorado, Linda Hogan, a member of the Chickasaw Nation. Her latest books have focused on the lives of animals and plants and have given me some comfort. Linda writes: "When I consider how we have unbalanced our world, changed its tilt by damning our waters, we have created the opposite of creation. Because of the many changes, another part of what we are charged with it to remember. Writer Meridel Le Sueur called it re-membering the dismembered.

Re-member. I want to remember this land, remember that plants are alive and speak with each other."

Bears Ears is a good place for re-membering, and not just because of its long Native history. All weekend long I find myself writing down names in my journal. Names of the people I meet, sure, but also names of plants and trees and birds. I know this will not stop climate change. But it feels right. If there is a way forward, and I'm not sure there is, it is toward nature.

It is so hard to imagine beyond the self, let alone imagine globally. My smallness is showing here. Which in a way is my theme. We are all small. It is hard being large. The extinction of the world may concern us, but our smaller selves are so much more interesting. Our demons can't ever really be exorcised. We need to carry on despite them. I know how small I am. I know that because I live with myself every day. I'm assuming that you, reading this, are equally small. That what we call largeness is an occasional, even spasmodic, effort. An effort that mostly fails. But maybe sometimes doesn't.

For us there is only the trying.

ONE LAST TRIP

I wake in a fog at my mother-in-law's house in northern New Jersey. Outside the window I can see the Hudson and that vulnerable city, New York, where we have just spent two days. My groggy mind starts to plan out the trip back south to Carolina: where I will put the paintings my mother-in-law gave Nina, and whether we should try to make the drive in one day. And then I have the sudden acute sense that I have forgotten something. That I have left something behind.

Oh right. I have. Our daughter.

Hadley will not be returning south with us. We will be leaving her in New York City, a college freshman.

How strange to have left her behind. Behind in this big uncertain world.

I expected Nina to cry when we hugged goodbye in front of Hadley's dorm but she didn't and neither did I. It was Amber, Nina's cousin, who was watching the parting from ten feet away who ended up in tears. Over the next week out own lives will be in the shadows. We will live only vicariously, following our child through texts and calls as she goes to her first class, meets her roommates, makes friends, and attends something called Drag Bingo in Washington Square Park. We will do our best to ignore the news that on Hadley's first day of classes a 25 year old woman, not a student, has been shot and killed in front of University Hall dorm at NYU.

Sometimes ignoring is all you have.

Not long before we made the trip to New York Hadley and I went for a swim near our home on Wrightsville Beach. We sat with our toes in the water as the tide swallowed most of the beach at the island's south end.

Not long before we made the trip to New York Hadley and I went for a swim on Wrightsville Beach. We sat with our toes in the water as the tide swallowed most of the beach at the island's south end.

"Climate change is not at the forefront of my brain," my daughter said when I ask about the rising water. "I am still a nineteen-year old girl. But when I am alone and really think about it it freaks me out."

Maybe she, like 97% of the population, like you perhaps, is sick of hearing about the climate crisis. Maybe she is ready for her father to start a different book on a different happier subject.

She pointed at the water.

"But it's here. This is not something that is coming. Even if I didn't have an overactive imagination I could see it. You can see it right in front of you. This is my childhood that is going under water."

The sentiment is a poetic one but the facts back her up. Many years ago, when I first started studying storms and traveling with Orrin Pilkey, I paid a visit to the M.I.T. offices of Kerry Emanuel, a professor emeritus of atmospheric science and one of the country's leading climate authorities. Back then many thought the jury was still out on whether or not warmer waters would lead to more intense storms. Emanuel assured me that what common sense suggested was true: hotter waters lead to more violent hurricanes.

"We have a heavily subsidized coastline," he told me then. "Subsidized by a corrupt insurance industry." He described how the insurance industry allowed people to build next to the shore without taking the financial risk. How someone living inland might pay as much in taxes as someone living on the shore.

Then, as I was leaving, after the formal interview was over, he said something that has stayed with me.

He described what he called the historical "natural human ecology" of the coast.

"The natural human ecology of the coastlines tends to be a few castles or mansions built very solidly that will withstand anything nature has to throw. But only a few—everything else is sea shanties. These shanties or cottages are disposable and people don't put anything of value in them and don't insure them. Every now and then they get wiped out and that's to be expected. It's the same all over the world…it's very democratic."

In other words, living by the shore had always meant taking a large risk. Taking a shot.

Sitting on Wrightsville Beach, Hadley and I were surrounded by huge homes that strove to be castles but that were still shanties at heart.

After Hadley left for college I wrote to Emanuel and asked him for a prediction for when my daughter was sixty. It had been more than a decade since we last spoke, and during that time his earlier predictions had borne themselves out.

He wrote that in his opinion "the most dramatic changes one might see by 2063 will be worse flooding by a combination of stronger storm surges and much increase in rainfall. We also expect wind speeds in the strongest storms to be up to 20% higher."

He also pointed me to a recent paper that he had co-authored with Avantika Gori, Ning Lin, and Dazhi Xi of Princeton, in which they had created models for the future of the coast. As it happened, one of the cities they had modelled was my own.

The takeaway of the study was that climate change greatly exacerbates the combined threat of extreme rainfall and storm surge during storms. During Florence, Wilmington had seen thirty inches of rain, and the models suggested that this might be the future for my adopted hometown, with predictions of overall rainfall during storms increasing by 32%, and storms lingering longer, just as Florence and Sandy had done. Emanuel and his colleagues used relatively modest sea level projections, nothing close to Hal Wanless or Orrin's predictions, but still found that the hazards of sea level rise and increased rain will likely lead to massive coastal flooding. Storm surges that will strafe across the island where Hadley and I were sitting. A blunter version of the report might read: the angry drunk is coming.

What the report described was a climate that felt like the tropical rainforest my home is well on the way to becoming. We already have mosquitoes in February. Around the world over the next forty years we can expect southern climates to migrate north.

"Let's get out of here," I said to Hadley.

"Let's."

He stood up, brushed off the sand, and walked back toward the dunes.

I'd had my doom dose for the day.

I am not a climate denier but I do have sympathy for some sorts of denial.

I understand that not thinking about a thing might seem to provide protection from it. And the way the immediate trumps the long term. Several jolts that have little to do with world-wide crises have begun to run like tremors through my small life. A recent blood clot in my leg made me more anxious than any melting glacier, and my mother's condition continues to worsen. Now the fact that we are leaving Hadley in New York looms larger than the possibility that that the same city might be underwater in forty-two years.

And so I end without concluding. I have told you what I have seen, what I have witnessed, as I traveled around this country. But I have no conclusions to offer. I can't pretend I know exactly where we are going. Like everyone else, I am caught in the amber of my own life.

Hadley is out in the world now. I hope she takes care. And I hope we take care of it, the world, so that the planet she finds as she grows older is not the one of the most dire predictions.

I don't know if we will do that or not. I don't know what will happen next.

— — — —

ABOUT THE AUTHOR

For twenty-five years David Gessner has reported from climate hotspots, from the Gulf of Mexico during the BP oil spill, to fracking towns and fires in the West, to the fragile Outer Banks where homes are being swallowed by the seas. He has been recognized for changing the face of nature writing, both in his own work and through the magazine he founded, *Ecotone*. The Washington Post writes: "For nature-writing enthusiasts, Gessner needs no introduction. His books and essays have in many ways redefined what it means to write about the natural world, coaxing the genre from a staid, sometimes wonky practice to one that is lively and often raucous."

Gessner is the author of twelve books that blend a love of nature, humor, memoir, and environmentalism, including the New York Times bestseller, *All the Wild That Remains*, and his latest, *Quiet Desperation, Savage Delight: Sheltering with Thoreau in the Age of Crisis* and *Leave It As It Is: A Journey Through Theodore Roosevelt's American Wilderness*. A professor at the University of North Carolina Wilmington, his magazine publications include pieces in the *New York Times Magazine, Outside, Sierra, American Scholar, Audubon, Orion*, and many other magazines, and his prizes include a Pushcart Prize and the John Burroughs Award for Best Nature Essay for his essay "Learning to Surf." He has also won the Association for Study of Literature and the Environment's award for best book of creative writing, and the Reed Award for Best Book on the Southern Environment. In 2017 he hosted the National Geographic Explorer show, "The Call of the Wild."

He is married to the novelist Nina de Gramont, whose latest book is *The Christie Affair*.

TORREY HOUSE PRESS

Torrey House Press publishes books at the intersection of the literary arts and environmental advocacy. THP authors explore the diversity of human experiences and relationships with place. THP books create conversations about issues that concern the American West, landscape, literature, and the future of our ever-changing planet, inspiring action toward a more just world.

We believe that lively, contemporary literature is at the cutting edge of social change. We seek to inform, expand, and reshape the dialogue on environmental justice and stewardship for the natural world by elevating literary excellence from diverse voices.

Visit www.torreyhouse.org for reading group discussion guides, author interviews, and more.

As a 501(c)(3) nonprofit publisher, our work is made possible by generous donations from readers like you.

Torrey House Press is supported by Back of Beyond Books, the King's English Bookshop, Maria's Bookshop, the Jeffrey S. & Helen H. Cardon Foundation, the Sam & Diane Stewart Family Foundation, the Literary Arts Emergency Fund, the Mellon Foundation, the Barker Foundation, Diana Allison, Karin Anderson, Klaus Bielefeldt, Joe Breddan, Casady Henry, Laurie Hilyer, Susan Markley, Kitty Swenson, Shelby Tisdale, Kirtly Parker Jones, Katie Pearce, Molly Swonger, Robert Aagard & Camille Bailey Aagard, Kif Augustine Adams & Stirling Adams, Rose Chilcoat & Mark Franklin, Jerome Cooney & Laura Storjohann, Linc Cornell & Lois Cornell, Susan Cushman & Charlie Quimby, Kathleen Metcalf & Peter Metcalf, Betsy Gaines Quammen & David Quammen, the Utah Division of Arts & Museums, Utah Humanities, the National Endowment for the Humanities, the National Endowment for the Arts, the Salt Lake City Arts Council, the Utah Governor's Office of Economic Development, and Salt Lake County Zoo, Arts & Parks. Our thanks to individual donors, members, and the Torrey House Press board of directors for their valued support.

Join the Torrey House Press family and give today at www.torreyhouse.org/give.